D0913960

Walls

Walls

Enclosure and Ethics in the Modern Landscape

THOMAS OLES

The University of Chicago Press   Chicago and London

THOMAS OLES has taught and practiced landscape architecture in the Netherlands, United Kingdom, and United States. He is the author of *Go with Me: 50 Steps to Landscape Thinking.*

The University of Chicago Press, Chicago 60637
The University of Chicago Press, Ltd., London

Printed in the United States of America

24 23 22 21 20 19 18 17 16 15   1 2 3 4 5

ISBN-13: 978-0-226-19924-5 (cloth)
ISBN-13: 978-0-226-19938-2 (e-book)

DOI: 10.7208/chicago/9780226199382.001.0001

Library of Congress Cataloging-in-Publication Data

Oles, Thomas, author.
Walls: enclosure and ethics in the modern landscape / Thomas Oles.
pages; cm
Includes bibliographical references and index.
ISBN 978-0-226-19924-5 (cloth: alk. paper)—ISBN 978-0-226-19938-2 (e-book) 1. Walls. 2. Walls—Social aspects. 3. Walls—Moral and ethical aspects. 4. Landscapes—Philosophy. I. Title.
NA2940.O44 2015
721'.2—dc23   2014017602

♾ This paper meets the requirements of ANSI/NISO Z39.48-1992 (Permanence of Paper).

For my children

The city being new and hitherto uninhabited, care ought to be taken of all the buildings, and the manner of building each of them, and also of the temples and walls.

PLATO, *LAWS*, 6

No. Leave the wall.
Remember—
You must always leave the wall.

THE FANTASTICKS

Contents

Acknowledgments ix

Prologue xiii

1 Good Fences, Bad Walls 1

2 What Walls Were 20

3 Constructions of Sovereignty 66

4 Recovering the Wall 114

5 Toward an Ethics 154

Epilogue 173 *Notes 185* *Index 211*

Acknowledgments

Any book, but especially a first book, and even more a first book on a vast and unruly subject, is at once a journey, a conversation, a lesson, an obsession, an affliction, a flirtation with the abyss, and a flight to realms, in the world and in the self, that one could not, on starting, have imagined existed. And at every crossroads along this tortuous way were the people—teachers and guides, readers and critics, friends and family—without whose example, support, and forbearance this volume would have fallen as flat as many of the walls described in its pages. They are truly my coauthors, and I am forever in their debt. It is impossible to name all these people here. But let what follows serve as a first reckoning, and a lasting testament to my gratitude.

I have had, over the long course of this project, the privilege to study with three titanic intelligences who remain the standard I set myself as a scholar and teacher. Leo Marx was the best kind of academic mentor, one who embodied in his very being that vanishing phenomenon, the public intellectual. His insatiable curiosity, in no way vitiated by age, munificence with his time and ideas, and impatience with the evasions and pomposity of academic language continue to inspire me with wonder and admiration. Rosalind Williams brought intellectual ballast and practical wisdom to the project when it was little more than an airy prospectus. She taught me that people, not ideas, have agency, prodded me gently but insistently to replace bluster with rigor, and reminded me of the uncomfortable truth that "at some point, to make a nuanced argument, you simply have to know more." Finally, Richard Sennett showed me by his own peerless example the way, through the act of writing itself, to marry a life of the mind with allegiance to justice. To all of them, my sincere thanks.

Many more individuals assisted me at important junctures, providing examples, suggesting interpretations, and offering criticisms that made the final manuscript inestimably richer. Eran Ben-Joseph and Julian Beinart in the MIT Department of Urban Studies and Planning were inexhaustible founts of both specialized knowledge and common sense. Jeremy Bunn inspired the beginning of the prologue, while Clifford Frasier discussed with me at length, on a glorious fall afternoon in the dunes of North Holland, the example that opens chapter 2. Robert Fogelson, John Harbison, Thomas Shou, Constance Guardi, and Syed Sikander Mehdi all gave generously of their time and expertise to answer questions about individual cases. Julie Hansen of the University of Uppsala made a characteristically meticulous reading of the draft manuscript at a moment when I needed it most. Marieke Timmermans, former head of the Department of Landscape Architecture at the Academy of Architecture in the Amsterdam School of the Arts, did me the honor of inviting me to present two chapters as part of that institution's distinguished Capita Selecta lecture series.

I have also profited, less directly but no less greatly, from all those mentors, colleagues, and friends who have shaped, over the past decade, my vision of an engaged landscape scholarship for a wide audience. Kenneth Olwig of the Swedish University of Agricultural Sciences; Jørgen Primdahl of the University of Copenhagen; John Forester and Paula Horrigan of Cornell University; Lizabeth Cohen of Harvard University; Jeff Hou, David Streatfield, and the late Harold Swayze of the University of Washington; Gail Dubrow of the University of Minnesota; and Kristina Hill of the University of California at Berkeley all contributed, each in his or her own inimitable way, to making me the scholar I am today. I am grateful for their continued example and support, both professional and personal. I owe a special debt in this regard to Shelley Egoz, professor in placemaking at the Norwegian University of Life Sciences, for organizing, along with Jala Makhzoumi and Gloria Pungetti, the "Right to Landscape" workshop at the University of Cambridge, where in 2008 I had the privilege to present an early version of this project. Many of the people I met at that stimulating and memorable event have become regular collaborators, trusted readers, and close friends.

The argument of this book is made partly through images, and many individuals helped guide me through the maze of copyright law that represents, particularly for the novice, a bewildering parallel universe to the writing of the manuscript itself. Among these were several who extended their hands in ways that went well beyond the merely transactional. Dean MacCannell of the University of California at Davis, Alex McCoskrie of Hadrian's Wall Trust, Marcus Harpur of Harpur Garden Images, Carlos Quijano of the City of Miami Shores, Florida, and the photographers Sue Sinton Smith and Derek Harper all offered their time and effort to secure images whose absence from these pages would have been sorely felt. My research assistants, Ryan Wright and Andrea Haynes, were indefatigable in tracking down copyright holders all over the globe, rising again and again at odd hours to call Denmark or Israel. Finally, I want to acknowledge the gifted artist Ida Pedersen, whose distinctive pen-and-ink drawings were commissioned for this book. They bring lost worlds to life again.

The University of Chicago Press has eased this outsider into the world of publishing with invariable solicitude and good humor. I want to thank my first editor, Robert Devens, now of the University of Texas Press, for taking on the project, then continuing to support it through repeated career changes, international moves, and missed deadlines. Russell Damian negotiated with great aplomb the editorial interregnum that coincided with my completion of the manuscript. My current editor, Tim Mennel, has been a model of sobriety and acumen, reading drafts multiple times and offering essential suggestions on matters of tone, subject, and organization. Nora Devlin, despite her recent arrival at the Press, has shepherded the manuscript into production with deftness. Finally, I thank the three outside readers who took the time to carefully read the entire text at two stages of writing. Their extensive and incisive comments, particularly those given as the manuscript underwent an agonizing metamorphosis from its larval stage, were indispensable. This book would have been far the poorer without them.

But there is one extraordinary individual who merits a paragraph all her own. Anne Whiston Spirn began by sparking my earliest interest in landscape architecture with her book *The Granite Garden,* went on to advise me as a doctoral candidate at MIT, and has now become, to my inestimable pleasure, both a vital critic and a dear friend. There is no single person whose influence on this project has been greater over the six years of its conception, gestation, and birth. Anne's ecumenical vision of landscape, skill and discipline as a writer, generosity with her time and ideas, and infallible good sense quite simply made *Walls* the book it is—or a book at all. For this I extend my deepest gratitude and admiration.

Acknowledgments are a tricky business. One needs to save the biggest people for the end, for the ends of things, one learns by experience, are what is borne away. Yet there is no question who belongs here. My mother, poet Carole Simmons Oles, taught me how to write. My father, architect Paul Stevenson Oles, taught me how to see. Most of all my spouse, philosopher Adele Lebano, who after inspiring many of its ideas, tolerated this project far longer, and with far greater equanimity, than any person should have been asked to do, taught me how to think—and then, what is harder, how to stop. To her I send all my love and all my thanks.

The very biggest people in this list, though, are the very smallest ones. I dedicate this book to my daughter Beatrice, whose morning steps and squeals I hear above me at this moment, and to our second child, nameless yet but already with us, beating and waiting in a walled, watery world. They are the reason I write, and the reason I stop.

Prologue

Once, toward the end of my studies to become a landscape architect, I visited a site with a friend I was working with on a design project. Neither of us owned a car, so he suggested we take the bus to the site, on the edge of a sprawling western city. As is common with suburban bus service, the trip involved transfer after transfer, baffling combinations of zones and tickets. At one transfer point we had to wait over an hour for a bus that would take us deeper into the foothills of the mountains. We got off and found ourselves at the crossroads of two major highways, amid fast-food restaurants, mini-marts, half-empty parking lots, and the tangles of telephone and power lines that make up the landscape of modern exurban America.

We huddled under the narrow bus shelter in silence, doing our best to avoid the April drizzle dripping from the roof. I was about to pull out an old paperback of H. G. Wells I carried for situations like this when my friend suddenly spoke.

"Look over there," he said, gesturing broadly with his left hand, a bit too cheery for the weather, I thought. "While we're waiting for the bus, let's do a little design exercise. Consider this: What's the cheapest, quickest, least complicated thing you could do to improve this place? Don't think too much about it. Don't analyze. There's a very simple answer, and it's right under your nose."

I squinted at the scene, trying to keep the rain out of my eyes. In front of us was a slick four-lane road where cars and trucks were gaining speed, perilously close to dousing us with muddy water as they passed. Behind us stood the windowless wall of a restaurant, to either side assorted buildings and shacks that looked no more than twenty years old and designed to last no longer than ten more. On all sides winding roads led up from the main roads to subdivisions hidden on the forested slopes above. To me the place seemed all but irredeemable. I shrugged, defeated and waiting for a hint.

"Look behind you again," my friend said. "That wall. If they punched a hole and started to sell coffee from a pass-through, then turned this bus shelter around to face the wall and added a few chairs, it would transform the experience of the commuters who have to wait in this place day after day. It would cost practically nothing, would increase the restaurant's business, and maybe even encourage the use of buses."

"But that's completely utopian!" I protested. "Something like that wouldn't happen anywhere."

"You have to imagine the possibilities," my friend answered evenly. "You have to see normal, everyday things in a different way."

I looked at the wall, made of white concrete blocks. It had simply faded into the whole bleak scene; indeed, it had seemed part of the problem. I tried to imagine the world my friend had just described. On a wet morning like this one, I could see a small crowd gathered around the window in the wall, a line forming as people waited for their buses. People talked to friends, chatted with strangers, and wondered what had inspired the restaurant owners to open this peculiar coffee stand. As people got their coffee, steaming in the damp air, they made their way across the sidewalk to take seats in the bus shelter, now open on both sides. Because the spot was so popular, the restaurant owners had arranged plastic chairs along the wall and inside the shelter . . .

Envisioning the scene, I saw that my friend was right. A small change in the way someone thought about the wall would have begun to redeem its environment. Through a relatively modest effort, recovering that wall would have begun to create, out of no place, a place.

Shortly after this episode, I moved to Denmark to continue my studies of landscape architecture. I expected to find a socialist utopia of solidarity and fairness. The confluence in Denmark of the welfare state and the tradition of modern design that had given the world Arne Jacobsen and Jørn Utzon sealed my notion that this was the best of all possible places. Yet when I arrived, the news was bad. I found a safe, modern, prosperous society in the throes of xenophobia. A right-wing government had come to power and nearly halted all immigration to Denmark.

The new politics was couched in the language of cultural triumphalism, the sense that Denmark was a special precinct reserved for the elect, who did not include people with dark skin or thin wallets. These changes in the political landscape came to shape the way I looked at the physical landscape. As I walked through cities, towns, and fields, a stranger in a strange land, I began to notice something distinctive about the Danish environment, something that at first had eluded my attention. This was the great variety of physical enclosures in the landscape. From traditional churchyards where each plot was surrounded by its own miniature hedge, to the hollowed-out perimeter blocks of Copenhagen, to the endless suburbs made up of one hedge-enclosed lot after another, each concealing a comfortable modern house, enclosure seemed to be everywhere—including, ultimately, in the islands and peninsulas of Denmark itself.

I began to wonder about the convergence of the political and the material. On the one hand, it seemed that all this enclosure symbolized some kind of cultural closure; on the other, the ubiquitous walls and fences seemed to express the egalitarian and democratic character of Danish society, with each resident allotted a small, calm place of refuge and peace. The designer in me could not help admiring the results in the landscape, which was extremely legible and orderly, but the liberal in me wondered how they related to the nascent atmosphere in the country. Certainly the relation was anything but simple. While it was risky to read in all that physical enclosure the expression of a desire to exclude an elusive "other," it seemed equally unlikely that the coincidence was entirely fortuitous. What was clear, as I learned from talking to people on my many walks, was that what I was seeing was almost invisible to the Danes themselves. When I asked about all this enclosure, people would often look at me as if I were mad. Later I would learn that these habits of enclosure were not somehow endemic to Danish culture but had been taught and learned in the 1960s, when home and garden magazines encouraged readers to enclose yards with high hedges and fences. The lesson had been absorbed so well that it was no longer subject to critical reflection. In Denmark a landscape of enclosure simply "was" in much the same way that unfenced lawns are seen as natural and inevitable in America.

When I returned to the United States and began to practice as a landscape architect, I was eager to reproduce the physical attributes of Danish enclosure in my work without replicating their cultural and political associations. In particular, I wanted to test whether it was possible to enclose a parcel of land in the American suburban landscape so that the fence around it at the same time provided a sense of enclosure and safety, heightened the legibility of the public landscape, and increased rather than limited social contact. In the words of sociologist Peter Marcuse, I wanted to make walls without creating boundaries.[1] What were the impediments to building such walls in a landscape very different from that of tiny Denmark, in a large and complex society where the difference between parcels could easily mean the difference between cultures, languages, or entire value systems?

In Denmark I had grown fascinated by the practice of weaving wattle from young willow shoots. Though wattle is common throughout northern Europe, I had rarely seen it used in the United States. I began to experiment with different types of wattle made of willow, red alder, and maple, creating variously dense or loose meshes according to the properties and age of each species, some supple and forgiving, some hard and brittle. I performed these experiments in my own backyard, mocking up the different materials for friends and acquaintances to judge.

When I was satisfied that I had found a suitable technique and species mix, I drafted a friend and his old Toyota pickup, and we drove around salvaging windfall from the streets near my house. It had been a stormy autumn, and there was no shortage; city workers had piled branches on the margins of paths, fields, sidewalks, and vacant lots. We helped ourselves, loading the slender branches of ash, maple, and red alder onto the truck as curious passersby stared. After dumping several truckloads in my yard, we set the posts of the fence and connected each pair with three cedar cross members. Then, over several long afternoons, we rammed branch after branch into the wet earth and wove them carefully between the members. As we worked, we grew attuned to the particular properties of each wood, from the flaky gray bark of the alder to the rigid, shiny shoots of the bigleaf maple. We noticed, too, that the interlocking branches had begun to attract birds. Thrushes and bushtits flew in from all directions and lighted in the upper reaches of the fence, gazing at us as we worked. This unexpected benediction led us to abandon our plan to cut the wattle down to a uniform height. Some weeks later, the lower branches even began to sprout leaves. Without our intention, it had turned into a living fence.

But it was the human relationships around the fence that surprised me most. It was as though the care we put into its making produced a friction between inside and outside; people stopped, took notice, and talked to us as they made their way down the street. Later I recorded

1 Wall without a boundary: living fence, Seattle, 2003. Courtesy of Eric Gould.

reactions from the kitchen window. Invariably people slowed down, looked, and talked when they drew alongside the strange construction. They ran their hands along the shoots, craning their necks to see into its upper reaches. The fence seemed to make people aware of their environment. It estranged at the same time as it oriented. It thickened relationships and practices, both human and animal.

One could argue that this test and its results were in no way "scientific." This part of the city was not diverse; the people passing my lot were generally white and middle class. How different would the results have been otherwise? Also, the final fate of the fence suggested that not everyone shared my ethical and aesthetic preoccupations; returning some years after I had sold the house, I noticed that my boundary had been replaced by an opaque board fence. But whatever its limitations, the experiment suggested to me the beginnings of an ethical investigation, one that would conceive of walls and fences as objects that stitch the landscape together and would claim these objects as sites of reflection and speculation.

Several years later, I was project leader for the landscaping of a real estate development in an American city with a history of violent crime and racial tension. The site was an abandoned Catholic hospital that occupied two city blocks in one of the worst neighborhoods in town. The structures, which were of high architectural value, were to be converted into residential apartments and housing for senior citizens. The planners hoped the project would stabilize the neighborhood by lifting property values and incomes. The design team was good: a developer with a social conscience, an architect with a history of progressive projects in difficult social environments, and a landscape architect with similar values and experience.

The project was thus similar to many urban redevelopments, with one significant exception: the site was entirely enclosed by a seven-foot brick wall. The wall was part of the original hospital complex and therefore listed as a historic structure, so it could not be altered in any significant way. Thus a project sold to local planning officials as a seed for future neighborhood development was destined to be sealed off from its surroundings. The architects had planned only a single perforation, along the west side where an elevated train ran, for use when the predicted wave of gentrification broke. Until then, this aperture would be blocked by a heavy steel gate opened at the discretion of those inside.

As in Denmark, the designer in me was drawn to this wall. I saw in it an antidote to a certain lack of clarity and legibility in the American landscape. The wall was carefully built and clearly worth preserving. But once again, the very thing the designer in me valued, the social critic reviled. An emblem of exclusion, the wall seemed to undermine the premise of the entire project. This tension followed me throughout the design process, raising inconvenient questions. How might a nearly impermeable boundary mediate between people with money and a city of people without it? How could a walled compound possibly be justified in a context where gated communities are most often associated with the decay of community life? How might the mandated presence of a wall throw into doubt the legitimacy of historic preservation itself? And finally, what were our obligations, as designers and as citizens, to consider these issues and respond through built form?

However one chose to answer these questions, it was the margin of the site, the place where the "work area" ended and the world began, that seemed to lie at the very center of the challenge the project posed. Yet in our office and at meetings with the owner and architect, that margin rarely came up in discussion. The wall's existence was either taken for granted or awkwardly skirted. Like the underground pipes or overhead wires leading into the site, it was just "there," part of the infrastructure of the city, hidden in full view, best ignored or forgotten.

For everyone involved, it seemed, the buildings within the wall were the essence of the project. But it was the wall, not those buildings, that would define the relation between the development and its environment. Yet the wall had no constituency, no adherents, no advocates. When, green as I was, I proposed making small holes in the wall to better connect the development to its surroundings, I was reminded to "concentrate my energy" (our time, the client's money) at the center of the site. The center was the place of economic and social value, and our interest lay in those who would live there and what they would pay. Somehow, when the time came the wall, and with it the world beyond, would simply take care of itself.

But walls do not take care of themselves. The shapes they take and the relationships they set in motion are in no way inevitable. Like the rest of the things people make, walls both reflect and create values. As designers we were presented with the challenge—the opportunity—to do the second. The reason we did not rise to that challenge, I believe, was that we lacked ethical standards for judging the appropriateness of the wall as it was or might become, for assessing its real and potential performance in the conditions we had inherited. Unlike the general rule that says a house needs a roof, and that a house without a roof is unlikely to be satisfactory, there was simply no standard we could turn to that would have helped us understand what was good or bad about the wall. It was not that we refused to engage the wall, then, but that we lacked the ethical apparatus for doing so. We did what most people would do: repair to the center, never exploring how the wall might both separate the site from its surroundings and stage a productive relationship with them.

—

The boundary is one of the idées fixes of contemporary life. Boundaries of the self, boundaries of politics, boundaries of nations and childhood and disciplines populate the pages of academic journals, the press, and the Internet. People speak of breaking through boundaries of culture, erasing boundaries to understanding, crossing boundaries between professions, challenging boundaries of gender. Wherever one looks, it seems, boundaries are being built and being broken. Yet despite all this preoccupation with boundaries, somehow a fundamental reality has been lost. It is this: boundaries are not just ideas, not just metaphors or images, but real things of wood and stone, metal and earth. Boundaries, in short, are objects made and maintained by people.

As the tales above suggest, making these objects presents a whole range of ethical and moral problems. What is the nature of our obligation when we construct a boundary? How can we bound and at the same time be generous? What is the relation between some people's right to enclose themselves and the rights of others who find themselves excluded? How can boundaries function as places where social relationships are nurtured rather than suppressed? To use Marcuse's turn of phrase, how can we begin to build walls of support rather than walls of fear?

Because people lack ethical standards for thinking about how walls should perform, they tend to confront these questions piecemeal—when they confront them at all. Walls, fences, ditches, and hedges may be the warp and woof of the landscape, but their making has faded into the background of concern. They have become sites of forgetfulness rather than reflection. My own inability to answer my friend's simple question that rainy day was a function of that forgetfulness. I was, quite simply, blind to the potential of the wall as a place of truck—of interaction and exchange. In design practice, we were incapable of thinking about the wall around the hospital as anything other than an annoyance. And in Denmark, people were so accustomed to making hedges and fences in one particular way that they no longer even saw them.

This forgetfulness has created the moral vacuum in which bad walls are built and sustained. Across the world, it is not hard to find examples of such bad walls—walls that divide, separate, control, and exert the power of the strong over the weak. When one looks at the landscapes of the greatest injustice, whether Cold War Berlin or the contemporary West Bank, it is more than likely that one will find a wall running through them. Bad walls are the reason many people now see boundaries as irremediably unjust parts of any environment. They are the reason Robert Frost could center one of his most famous poems on the lines "Something there is that doesn't love a wall / That wants it down."

And yet. People have always made walls, and they always will. I suggest that this is not something to be lamented or resisted, but rather an opportunity to be seized and an obligation to be met. A great many walls in the world bear out Frost's words. We must use all the force in our hands and all the voice in our throats to condemn these walls and to demand their destruction when they should not stand. But we must do more than that. We must begin to conceive an ethics for shaping the walls that do exist and imagining those that might exist. In the modern landscape, it is not enough simply to want walls down.

We must learn to love them too.

ONE

Good Fences, Bad Walls

"Good fences make good neighbors. Fences don't make bad neighbors." This is what one United States senator said in spring 2006, justifying the amendment he had recently submitted to a comprehensive immigration reform bill. The amendment ordered the construction of 370 miles of triple-layer fencing, along with roads, lights, and sensors, on the boundary between the United States and Mexico. "Go to the San Diego border and talk with the people," he told the chamber. "There was lawlessness, drug dealing, gangs, and economic depression on both sides of the border. When they built the fence and brought that border under control, the economy on both sides of the fence blossomed, crime has fallen, and it is an entirely different place and a much better place. That is just the way it is."

Not all the senator's colleagues were swayed by this reasoning. One dismissed the fence as a "symbol for the right wing in American politics," while another condemned its "'fortress America' approach to real world problems."[1] Immigration reform died in the House of Representatives several months later, but a new bill was put forward the following autumn. It more than doubled the length of the fence as originally proposed; one section would extend, unbroken, nearly four hundred miles.[2] That bill passed both houses of Congress by wide margins.[3]

The senator's rhetoric was typical of the moment. The phrase "good fences make good neighbors" was invoked again and again during the summer of 2006, a mantra of national enclosure chanted by politicians, the media, and ordinary citizens. One of the senator's colleagues in the House told that chamber: "I recently returned from a week-long trip to the Mexican-California border, and I am convinced of one thing. Good fences make good neighbors."[4] "People may not always like fences," an engineering magazine instructed its audience, "but good fences make good neighbors, as other nations around the world are realizing."[5] A reader of the *Miami Herald* wrote to "remind people who think that closing our borders is cruel that 'good fences make good neighbors.'"[6] There was remarkably little difference, it seemed, between steel bollards and barbed wire separating two countries and an old hedge pruned by amiable suburbanites over the years.

These comments were mere drops in a global ocean. The phrase "good fences make good neighbors" routinely appears in a wide array of rhetorical contexts. In recent years it has been used to argue for marking the border between China and India, and to justify Australian naval maneuvers along that country's maritime boundary with Indonesia. It appeared, in English, in an Albanian newspaper detailing the state of that country's border with Greece. It was employed to explain relations between Singapore and Malaysia; the ethnic conflict in Kosovo; the Israeli occupation of the West Bank; and even fiscal policy in the United Kingdom. And of course it appeared in reference to fencing disputes of every conceivable kind, from a "VIP enclave" in Pretoria, to a walled dormitory in Montreal, to the fence built by a celebrity politician around her yard in Alaska.[7] "Good fences make good neighbors" seems to be far more than a proverb. It has become one of the "metaphors we live by."[8]

The transformation into metaphor is very recent. Throughout most of its history, when public order and even survival depended on separating livestock from crops, "good fences make good neighbors" was an axiom that simply stated common sense. The proverb was first recorded in Blum's *Farmer's and Planter's Almanac*, published in North Carolina in 1850, but it was centuries old even then.[9] For his *Elegant Extracts* in 1797, English essayist and minister Vicesimus Knox translated the Spanish saying *una pared entre doz vezinos guarda más la amistad*, common throughout Spain and its colonies in the Middle Ages, as "a wall between both, best preserves friendship"; Ralph Waldo Emerson later transcribed the entry in his journal.[10] Two centuries before that, the Reverend Ezekiel Rogers had touched on "the business of the Bounds" in a letter to John Winthrop, first governor of the Massachusetts Bay Colony.

Rogers wrote, "I have thought, that a good fence helpeth to keepe peace between neighbors; but let us take heed that we make not a high stone wall, to keepe us from meeting."[11] A good fence, in Rogers's estimation, was one that kept fields free of errant livestock but did not prevent contact between the people who tended them.

If good fences were associated with social harmony and public order, bad ones were widely taken to mark their absence. "Poor fences ever tell a sad story," wrote the *Boston Cultivator* in 1851. "When we see a farm with fences all going to decay the conviction is irresistible that some shocking legend is connected with its history."[12] Emerson's friend Henry David Thoreau, in his own journal, defined "bad neighbors" as "they who suffer their neighbors' cattle to go at large because they don't want their ill will,—are afraid to anger them." He called people who failed to maintain their fences "abettors of the ill-doers."[13] Good fences, in short, equaled good citizenship.

Yet people who say "good fences make good neighbors" today are unlikely to have this long history in mind. Instead, they are quoting, wittingly or unwittingly, a poem that is barely a century old. Since its original publication in 1914, Robert Frost's "Mending Wall" has all but merged in the popular imagination with the phrase that forms its twenty-seventh and forty-fifth lines.[14] While this has assured the proverb a permanent place in American—and world—idiom, it has tended to reduce the *poem* to little more than a half-remembered apology for walls and fences everywhere. The words of one newspaper columnist during the debate over the United States border fence were symptomatic: " 'Fences make good neighbors,' the poet said," he told his readers.[15]

But he was wrong. The poet did not say, or mean, that at all.

Something There Is

New England soil is a bottomless reservoir of stone. In garden and park, yard and field, each spring thaw brings up a new supply of granite, gneiss, and schist, one infinitely small installment of the inheritance left by the Laurentide ice sheet as it made its retreat ten thousand years ago. Like their predecessors in all the rocky places of the world, New England farmers piled boulders into mile on mile of walls along fields, roads, and woods. The walls were continually being brought down by harsh winters and wandering livestock, and remaking them was one of the unavoidable, unceasing tasks of rural life. By the early nineteenth

2 Iconography of enclosure: rural landscape with stone wall, New England, about 1910. Courtesy of the Library of Congress.

century, large parts of New England had been deforested, the landscape transformed into a lattice of walls draped over gentle hills.

Within a few decades, however, farmers, drawn by fertile lands to the west, abandoned all their work. The woods returned, blanketing almost all the lands that had been cleared. Walls that once divided fields and enclosed paddocks now crisscross woods or decay in stands of maple and white pine behind housing developments, shopping malls, and vacant lots. Like church steeples and town commons, the New England stone wall has become an icon of a lost and better time, enshrined on calendars and computer screens worldwide.

"Mending Wall" is rooted in this history. At first its subject seems quaint, even banal: two rural neighbors meeting in the first days of spring to fill the gaps that have appeared in the stone wall between their two properties. The opening lines alert the reader, however, that this will be more than a threnody for the practices of a bygone age. The poem is supposed to be about repairing a specific wall, but it begins by evoking a mysterious force that contrives to break apart walls in general:

Something there is that doesn't love a wall,
That sends the frozen-ground-swell under it,
And spills the upper boulders in the sun,
And makes gaps even two can pass abreast. (lines 1–4)

The awkward rhythm and contorted syntax of these opening lines heighten the strangeness of the "something" in question, a force that conspires to undermine language as much as things. But there is nothing mystical about this "something." It is frost, and the pun seems to ally the poem's narrator and author with nature itself against this and every wall. But there are other wall breakers in these first lines, too, hunters who leave "not one stone on a stone" (7) in their efforts to drive rabbits out of hiding (9). Whatever natural or human agents lie behind the wall's disintegration, their existence is attested only by the gaps they leave, which the narrator and his neighbor discover when spring comes ("No one has seen them made or heard them made, / But at spring mending-time we find them there" [10–11]).

"Mending Wall" thus opens not with the strength of walls, but with their essential *fragility*. Solid granite, it turns out, is no match for a cold snap or a determined hunter; without constant care, walls will always fall into ruin. The rest of the poem depicts one of the rituals of such care, the two neighbors slowly walking the length of the wall, gathering and replacing the boulders that have fallen out of it:

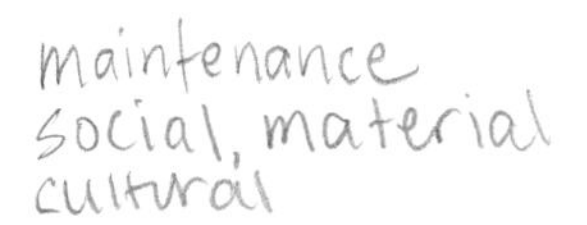

And on a day we meet to walk the line
And set the wall between us once again.
We keep the wall between us as we go.
To each the boulders that have fallen to each. (13–16)

These parallel, end-stopped lines are very different from the poem's opening. They suggest not instability, not fragility, but steady and purposeful labor; the men have performed this ritual many times and need not speak to each other to do it right. But if the narrator and his neighbor know their task equally well, they understand it in very different ways. As his neighbor works on in silence, the narrator begins to wonder why this wall is necessary at all in a place devoid of livestock, when neither of the two men is a farmer:

There where it is we do not need the wall:
He is all pine and I am apple orchard.
My apple trees will never get across
And eat the cones under his pines, I tell him. (23–26)

The neighbor answers this witticism with five simple words: "good fences make good neighbors" (27). The blunt syntax of this phrase contrasts sharply with the verbal contortions of the poem's opening lines and might close the matter on a colder day. But "spring is the mischief" (28) in the narrator, and he presses on, albeit now in his own head:

. . . I wonder
If I could put a notion in his head:
"*Why* do they make good neighbors? Isn't it
Where there are cows? But here there are no cows.
Before I built a wall I'd ask to know
What I was walling in or walling out,
And to whom I was like to give offense.
Something there is that doesn't love a wall,
That wants it down." I could say "Elves" to him,
But it's not elves exactly, and I'd rather
He said it for himself. (28–38)

This passage yields the second pun in the poem, even more outrageous than the first, and adds another agent, this time *super*natural, to the forces that want walls down. For the narrator, such play is all in the spirit of the thaw, mending wall "just another kind of outdoor game / One on a side" (21–22) in which the men must cast spells and issue threats to make the semi-animate stones remain in place ("Stay where you are until our backs are turned!" [19]).

But there is method in this play. The narrator wants his neighbor to justify the wall, to argue for its existence when the need that gave rise to it has vanished. He is not interested in swaying his neighbor through argument; he prods him so he will acknowledge "for himself" the absurdity of repairing the wall year after year. But the neighbor refuses to play this game. His only response is to quote again, because "he likes having thought of it so well," the words he has inherited from his forefathers: "good fences make good neighbors" (45). This obstinacy reduces him, in the poem's final lines, to a state of near barbarism in the narrator's eyes:

. . . I see him there
Bringing a stone grasped firmly by the top
In each hand, like an old-stone savage armed.
He moves in darkness as it seems to me,
Not of woods only and the shade of trees. (38–42)

Like the wall itself, the neighbor is a remnant of an older, darker world, one where words are only repeated, never invented. The narrator wants to flood this world with light, to break apart its walls of language and stone, but his efforts run afoul of the neighbor's proverb, the poem's last line. "Good fences make good neighbors" is thus not the moral of "Mending Wall." It is the wall that bars the two men, and the poem, from going any further.

And yet this cannot be quite right. It is certainly too easy to reduce "Mending Wall" to the neighbor's proverb, as is widely done. But it is equally mistaken to read it as a denunciation of every wall as "primitive, fearful, irrational and hostile," as Monarch Notes would have readers think.[16]

This becomes clear when the actions of the narrator are considered alongside his words. Despite his alliance with forces and beings that "want walls down," it is the *narrator*, not his neighbor, who initiates the mending ritual every spring ("I let my neighbor know beyond the hill" [12]), and it is he who fills the gaps left by hunters during the winter ("I have come after them and made repair" [6]). Not once in the poem does the narrator consider dismantling the wall or letting it disintegrate. Instead, he returns year after year to "walk the line." And while the neighbor might initially appear taciturn and, by poem's end, even prelinguistic, a closer reading suggests the narrator is the less communicative of the two men, his thoughts only one time emerging as speech. If only by repeating his favorite proverb, the neighbor does more to break the wall of silence between the two men.

If the wall embodies a darker, less civilized world, why does the narrator want to repair it? Why not simply let frost, hunters, elves break it apart? After all, there is no longer any practical need for the wall: cows no longer wander these fields, and apple trees do not leap across boundaries to eat pinecones. "There where it is we do not need the wall," the narrator reminds his neighbor (23). But this apparently straightforward statement begs an essential question. What, exactly, is a "need"? Saying that a wall no longer does one thing it once did is not at all the same as saying that it is no longer needed—and the narrator knows it. He returns to mend the wall not because it continues to serve some original purpose, but because for him the wall's utility, and its value, lies elsewhere.

The difference between the characters in "Mending Wall" is not their desire to mend the wall, then, but their reasons for doing so. For the neighbor, the wall is an object fixed in space much as a proverb is fixed in time, a guarantor of cordial, duly distant relations between the two men. It would be all to the good, no doubt, if the wall could continue to stand without human effort. For the narrator, however, the wall is something else entirely. It is less an object than a set of practices; less piled boulders than the walking beside them, the placing and replacing of one atop another every spring. For him the wall's value lies precisely in the fact that it *does* disintegrate. Its "function" is not, indeed perhaps never was, to separate livestock or define the boundary between two properties. It is to occasion the yearly ritual between the two men.

This is the heart of the narrator's grievance with his neighbor's proverb. It is not that the words "good fences make good neighbors" harbor no truth. It is that they are too easy, too automatic, too unthinking, that the neighbor will not "go behind" them, pull them down if only to put them up again for another year. It is this rebuilding, this breathing of new life into old forms of stone and language, that draws the narrator and makes him summon his neighbor year after year. It is no accident that the title of the poem is a gerund. Good fences do not make good neighbors; mending them does.

Finally, "Mending Wall" is less about two different men than about two parts of a single man. This reading is borne out by Frost's own words about the poem. "I've got a man there," he told one interviewer late in his life. "He's both of those people but he's man—both of them, he's a wall builder and a wall toppler. He makes boundaries and he breaks boundaries. That's man."[17] Later, asked about the omission of the poem's first line from a Soviet anthology, Frost said: "I could have done better for them, probably, for the generality, by saying: 'Something there is that doesn't love a wall, / Something there is that does.' Why didn't I say that? I didn't mean that. I meant to leave that until later in the poem." Frost called the first and last lines—"something there is that doesn't love a wall" and "good fences make good neighbors"—"detachable statements" that one must read together in order to fully understand the poem.[18] Thesis and antithesis, they express two impulses that coexist, often uneasily, in all people: the desire to enclose and the need to connect.[19] Every wall or fence implies its own destruction, the persistent "something" that "wants it down." But this something must be reconciled, again and again, with the equally potent thing in us that wants it up.

"Mending Wall" is about the way this reconciliation happens. Its second most famous passage ("Before I built a wall I'd ask to know") is an appeal not to tear down walls, not to ignore them, but rather to ask questions of them. The poem asks readers to scrutinize their own motives whenever they make or remake a boundary, to confront directly the tensions—between power and impotence, inclusion and exclusion, separation and connection—that every wall, every fence, every hedge or ditch embodies. Because these tensions can never be fully resolved, new and different questions must be asked year after year, wall after wall.

At root, then, "Mending Wall" is a poem about ethics. It is neither an endorsement of walls nor an argument for their destruction. It is a guide for making them better. And while its subject is a ritual from another time, its message is as important today as ever.

Segregation and Sovereignty

The collapse of European totalitarianism and the rise of electronic communication at the end of the twentieth century seemed to promise an age when boundaries of all kinds would fall into irrelevance. The political divisions that had emerged from the rubble of World War II were vestiges of a darker age; in their place, the "global village" first promised by

Marshall McLuhan in the 1960s was coming into view.[20] The American political scientist Francis Fukuyama confidently—and prematurely—announced "the end of history," and world media were full of similar portents.[21] "The world is growing increasingly integrated and unified," the *Jerusalem Post* confidently told its readers in November 1989, in a typical editorial about the events then unfolding in Berlin. "Its walls are falling. We must be a part of this new world order, open to it and the opportunities it provides."[22]

More than two decades later, large parts of this new order have been realized. There are now more democratic than undemocratic states. More countries lie within free trade areas than outside them.[23] National boundaries have virtually disappeared across Western Europe. Information and capital travel vast distances at the stroke of a key, bypassing oceans, checkpoints, and duties. People move across the world for work and leisure in increasing numbers, travel having long ceased to be a privilege of wealth. And most recently, each day sees another million or so users added to "virtual" social networks spanning languages, cultures, and continents. What geographer Manuel Castells famously called "the space of flows" is, for many people, not an abstraction but the space they inhabit every day.[24]

But it is a grave error to confuse the falling of legal or technological barriers with the disappearance of boundaries. Walls have fallen for some, but they have risen higher and higher for many others. There is a widening divide between those who move seamlessly amid Castells's "flows" and those who must contend with ever-tighter controls on their movements. Capital moves freely, but people do not, a rule that holds especially true for the poor and the desperate. Over the past decade, one developed country after another has raised legal barriers to asylum seekers and immigrants, and many have attempted to halt so-called economic migration altogether.

These new barriers are not only legal or procedural. As political scientist Wendy Brown has documented, more and more countries are backing up such restrictions with real fences and walls.[25] The 730 miles of fencing mandated by the Secure Fence Act of 2006 has now been built, much of it on private lands seized by eminent domain. The fence has split Mexican-American communities, impeded trade, and disrupted natural process along its entire length. A double chain-link fence ten feet high and topped with barbed wire surrounds the Spanish Moroccan enclaves of Ceuta and Melilla, designed to prevent Africans from entering the European Union. Botswana has constructed an electrified barbed-wire fence to block refugees from Zimbabwe, and India has built a chain-

3 Only one national wall: "border fence" between Naco, Arizona, and Naco, Sonora, in Mexico, 2006. Courtesy of Cameron Davidson/Corbis.

link and barbed-wire fence along much of its border with Pakistan.[26] In some cases these fences and walls are embedded in extensive no-access areas or "buffer zones," great gashes through what were once inhabited landscapes. The ironically named "Demilitarized Zone" that has run across the Korean Peninsula since the armistice of 1953 is the oldest and most extensive of these areas, but there are many others, from Kuwait to Cyprus to Kosovo.[27] Finally, since the early 2000s Israel has used an extensive system of high concrete walls and security fences to consolidate its sovereignty over the West Bank. The "seam line obstacle," as this system is officially called, has separated villages from their fields and wells, divided urban neighborhoods, and subjected Palestinian citizens to unprecedented restrictions on their physical movements. This elaborate apparatus of territorial control has been widely condemned by international organizations and was declared illegal by the International Court of Justice in 2004.[28]

These national walls are mirrored by the walls and fences that have risen in cities over the past forty years. In the United States, the idea of "defensible space," developed by the architect Oscar Newman in response to the riots of the late 1960s, led to the fencing and gating of hundreds of public streets during the 1970s and 1980s.[29] The trend intensified in the decades that followed, spreading from central cities

4 City as parcel: fenced municipal boundary of Miami Shores, Florida, 2008. Courtesy of Carlos Quijano.

to suburbs; it has been estimated that over 7 million Americans, or 6 percent of all households, now live in communities surrounded by walls or fences.[30] Some cities, such as the Miami suburb of Miami Shores, have gone so far as to fence their entire municipal perimeters.[31] But this is no longer just an American approach: from Russia to South America to the Middle East, walled enclaves, many defended by private security forces, are now common.[32] Architect Lindsay Bremner describes how barriers with cameras, sensors, and armed guards, built around wealthy neighborhoods of Johannesburg since the 1990s, have transformed large parts of that city into a "permanent frontier zone."[33] The walls scoring modern cities differ little from the barricades of nations; indeed, from Jerusalem to Brownsville, they are often the same thing.

The idea of a world without boundaries may seduce the imagination, but it is belied by evidence the body encounters at every step. Go outside—walk straight—and take note of the distance before a wall or fence blocks your path. Metaphorical walls may have fallen, but boundaries remain, before all else, *things* made by people. And their number in the landscape is not shrinking, but growing.

"To put up a fence," Rebecca Solnit has observed of the United States border fence, "is to suggest difference when there is none (though there will be), and to draw a border is much the same thing."[34] As far as it goes, this statement is beyond dispute. Boundaries do indeed create distinctions between people, groups, and places—distinctions that would not exist in the same form, if at all, in their absence. As both metaphors and things, boundaries cleave the world and make difference out of sameness.

But these words leave out as much as they include. For it is through difference that much of the material and social world comes into existence. Many religions share a similar creation myth in which an original being without contrast or distinction gives rise to the world in all its bewildering diversity. Polynesian myth, for example, tells the story of the supreme god Io, who brings forth all existence from primordial waters with the words, "Ye waters of Tai-Kama, be ye separate. Heavens, be formed!"[35] Plato, in the *Timaeus*, describes a "state devoid of reason or measure" before things were "marked out into shapes by means of forms and numbers"; the Greeks identified this state with the god Chaos, the first created being from whom all other deities sprang.[36]

Such myths parallel the development of human perception. Psychologist Jean Piaget demonstrated that one of the first ways infants construct their cognitive world is by learning to separate or "dissociate" elements in their visual field; "the more analytic perception becomes," Piaget wrote, "the more marked is the relationship of separation."[37] And English psychoanalyst Donald Winnicott noted how the process of individuation in children takes the form of a "limiting membrane" between inside and outside.[38] The self itself, in short, begins with a boundary.

Boundaries play an equally important role in the social and political world. Political philosopher Michael Walzer has called liberalism "a world of walls" marking out the distinct realms of state, market, church, and home.[39] Sociologists have long noted that strong boundaries can help to forge a sense of common purpose and mutual obligation within communities.[40] Some thinkers have gone even further, arguing that boundaries are equally necessary for extending concern toward people *outside* the groups they define. Boundaries in this view not only create solidarity with those who are similar; they also enable empathy toward those who are different.[41]

There are many boundaries in the world that separate what should be inseparable, divide what should be indivisible. Walls and fences break apart families, communities, and livelihoods, and make daily existence inconvenient, humiliating, or intolerable for millions. They project the power of the strong over the weak, the rich over the poor. They fragment ecology, impede movements of animals, disrupt the flow of water. Highways or greenbelts divide groups and races from one another; armed barricades around neighborhoods and parcels degrade ever more urban landscapes.

But it is a tenuous step from these statements about the ways walls are sometimes used to the claim that there is something essentially bad about walls. And yet this is the very step taken by a great many observers of the contemporary environment. In their understandable zeal to denounce bad walls, they often fail to distinguish between creating a *distinction* and committing an *injustice*. Consider, for example, the conclusion of *Fortress America*, a study of gated communities in the United States: "All of the walls of prejudice, ignorance, and economic and social inequality must come down before we can rendezvous with our democratic ideals. . . . Then the walls that separate our communities, block social contact, and weaken the social contract will also come down."[42]

As in the Monarch Notes reading of Frost, here walls embody "all that is primitive, fearful, irrational and hostile" in the people who build them. They are the inevitable result of prejudice, ignorance, and inequality.[43] When these evils vanish from the earth, the teleology of the passage promises, so, too, will the walls they have spawned.

While one can hardly take issue with its moral stance, this argument ignores history and defies common sense. The relation between the things people think and the things they make is never as simple as the passage above suggests. There is simply no evidence that walls per se are irreconcilable with democracy and community: there are many egalitarian, democratic societies where walls are widespread and many inequitable, undemocratic ones where they are rare. People have made walls in every imaginable condition, for innumerable reasons. If the modern landscape is full of bad walls, this surely bespeaks not some inherent badness of walls, but bad premises about what they can, and should, do.

In the *Nicomachean Ethics*, Aristotle defines goodness as the degree to which a given thing exhibits "virtue" (*arete*), or the characteristics that enable it to perform the "function" (*ergon*) for which it is suited. For example, the function of a knife is to cut; a knife with a sharp blade cuts well; therefore a knife's virtue, and its relative goodness, lies in the sharpness of its blade.[44] Walls are far more complex objects than knives, but it is common to think of their virtue in similar ways. This comes down to two fundamental assumptions about the functions of walls that, like the idea that knives are for cutting, are today widely seen as self-evident.

The first of these assumptions is that walls exist primarily not to *differentiate*, but to *segregate*. The distinction between the two may seem minor, but it is in fact great. "Differentiation" denotes any "change by which like things become unlike, or something homogeneous becomes heterogeneous"; it says nothing about keeping those things apart. "Segregation," by contrast, refers to "the isolation of a portion of a community or a body of persons" or "the separation of a portion or portions of a collective or complex unity from the rest."[45] It suggests not simply creation of difference, but prevention of exchange.

The idea that walls are meant to segregate people is widespread among both those who support them and those who oppose them. During the debate over the United States border fence, for example, many opponents warned that the fence would not work. "You show me a 50-foot wall and I'll show you a 51-foot ladder," one senator intoned, no doubt thinking such reasoning unassailable.[46] But these words suggest that their author shared with his adversaries a fundamental understanding of what it means for a fence to "work." For both, the "goodness" or "badness," and hence the justifiability, of the fence was a clear function of its capacity to segregate; the greater one believed this capacity to be, the more likely one was to support construction. The strongest argument opponents could muster against the border fence, in other words, was that it would be *permeable*.

The case of the United States border fence suggests a second, more insidious assumption: that the prime function of walls is to express and enforce sovereignty. It is not hard to see why this idea is widespread. The world offers up a nearly endless array of walls and fences made by one group to control or restrict the actions and experiences of another. Though these objects often provoke heated political debate, decisions about when, where, and how they get built ultimately derive from the power to, as political philosopher Carl Schmitt famously put it, "decide . . . on the exception."[47] Schmitt's characterization of sovereignty aptly describes the United States border fence, whose construction took place over the opposition of the Mexican government and thousands of American citizens, and after the abrogation of more than thirty federal laws.[48] Similarly, the "seam line obstacle" built by Israel deviates wildly from the so-called Green Line, the armistice line of 1949 and 1967, pushing deep into Palestinian territory and securing Israeli control over the West Bank.[49] The Israeli government has consistently justified this

5 Site of forgetfulness: sidewalk and yard before and after fencing, Cambridge, Massachusetts, 2008. Photographs by Thomas Oles.

route using the rhetoric of territorial sovereignty and cites its "legal right to build the Security Fence in order to protect the lives of its citizens."[50]

This statement and others like it portray walls and fences as weapons in a global contest among nations and groups for control over territory, people, and resources. But such walls account for only a small portion of the boundaries in the modern landscape. It is far more common for walls today to embody another, more familiar form of sovereignty: that of property. Freehold on a parcel of land often confers on the owner considerable rights to determine how that parcel will relate to its environment, whether the owner is a developer choosing the material for a wall around a gated community or a home owner pruning a hedge to four feet rather than seven. Such decisions affect the lives of many people in intimate and often profound ways, but they are almost always made without public deliberation; even when such deliberation does occur, private prerogative usually prevails over public interest. In other words, the objects marking property are today, for the most part, associated with claims of ownership rather than duties of citizenship.

Together these assumptions form a kind of ideology, or a collection of beliefs posing as facts. Walls and other boundaries, this ideology says, are unjust by nature; in Aristotle's terms, their virtue lies in exerting the power of strong over weak, rich over poor, propertied over landless. The highest purpose of any wall or fence is to prevent contact between people, groups, or nations. For this reason, the only good wall is a fallen one.

It is not that this account contains no truth; on the contrary, there is ample evidence in the landscape to support it. But its moral force is purchased at the cost of historical nuance. By suggesting that walls, like knives, can be reduced to a single virtue, it drains the long and complex story of boundaries of all its richness. It is blind to the fact that, as essential elements of houses, temples, fields, and cities for millennia, walls could never possibly be any one thing in the minds of the people and societies that make them. It is deaf to the fallacy of assuming that because walls are coercive in some situations, they have always been so and will always be so.

The result of this blindness is a landscape of walls commensurate with the limited expectations that people—both those who want them up and those who want them down—have of them. The very prevalence of such walls then further entrenches the faulty assumptions that gave rise to them in the first place. Like a chain of descent in the Old Testament, bad walls beget yet more bad walls. This alone would be reason enough to condemn them. But bad walls have another, more corrosive effect, one that goes beyond any single object or situation. They debase and delegitimize the essential human desire, and need, to bound. Bad walls make enclosure *itself* seem barbarous and perverse, an act fit only for "old stone savages," the result of little more than hardened ignorance or ancient prejudice. Bad walls make it altogether too easy to forget that, in most places and times, the acts, rituals, and objects associated with boundaries have been not socially destructive, but socially constructive. By obscuring what walls have done, bad walls impoverish aspirations for what walls can do.

Today, for the first time ever, more than half the human population lives in cities. In North America and Europe, urban residents already account for nearly eight out of ten residents; in Africa and Asia, the urban population has gone from one in ten residents in 1950 to over five in ten today. In China, it is estimated that 18 million people migrate from rural areas to cities every year, one of the most rapid urbanizations in history. The share of urban residents living in "megacities," urbanized areas of 10 million or more, has more than doubled since 1975; it is estimated that one in three of these residents live in areas classified by the United

Nations as slums. The urban population is expected to reach 5 billion by 2030, at which point the developing world will account for fully 80 percent of humanity.[51]

These changes will present immense technical, political, and moral challenges in the coming decades, as the geographical extent, ecological impact, and social and economic divisions of cities assume unprecedented dimensions. In many places cities already rival nations for political and economic power. But the urbanization of the world will also present challenges that touch on the more intimate realm of private action and individual behavior. One of these challenges derives from the reality of more people crowded into smaller areas. The urban landscape of the future will be divided into adjacent, sometimes overlapping, and often vastly unequal territories, something already evident in the cities of the developing world. In Bombay, for example, some 6 million people, or half the city's population, crowd onto only 8 percent of its land.[52] But increased density is not only a story of the poorest places of the planet. Lot sizes in North America, too, have been falling for three decades, and most American cities, after years of dispersion, are once again increasing in density.[53] Rich and poor, north and south, urban residents live, on the whole, in ever greater propinquity.

Such conditions will demand new ways of conceiving the relation between individual rights and social obligations. Inevitably this will mean addressing the way people deliberate and make boundaries. It will mean finding new ways to balance the legitimate human impulse to differentiate, define, and protect with the practical reality of life lived in proximity. In an urban world, it will no longer be a question (if it is even now) of once a year mending a distant wall "beyond the hill," but rather of negotiating the walls and fences outside the door, around the corner, down the street. These boundaries will mark, in ways increasingly difficult to ignore, not the difference between "inside" and "outside," but the difference between "my" or "our" inside and the insides of others.

Even in a just world, Frost knew, people would probably still build walls; that they will do so in an unjust, ever more crowded one is a near certainty. Yet the standards for building and judging walls that prevail today are not sufficient to meet the moral and practical challenge this presents. Should they be allowed to remain unquestioned, the faulty assumptions I have described virtually ensure that many of the walls that get built in the coming years will be neither just nor justifiable. These faulty assumptions must therefore be replaced by new and more rigorous standards that change the expectations people have of walls and the demands they make on them. Such standards will emerge, however, only when people begin to apply the same reflection and questioning to every boundary in their environment—both those that will be built and those that already stand—that Frost brought to an old stone wall north of Boston more than a century ago.

In the modern landscape—almost by definition a subdivided and urban landscape—the things that mark boundaries must do more than segregate people or express sovereignty. They must begin to reconcile, in deliberate ways, the tension between bounded space and shared citizenship. This will require far more than broadening ideas of what walls do.

It will mean imagining anew what walls are.

TWO

What Walls Were

The book of Joshua tells of two Israelite spies sent secretly out of the town of Shittim, northeast of the Dead Sea, to reconnoiter the lands west of the Jordan River. The men make their way to the important city of Jericho in order to determine the strength of its defenses. When they arrive, they seek shelter in an unusual place: the dwelling of a prostitute named Rahab, whose "house was upon the town wall."[1]

The king of Jericho, who has been warned of the spies' presence, summons Rahab and demands that she produce them. Rahab has hidden the men under stalks of flax on the roof of her house, but she tells the king she has seen them leave the city. She tells the spies that she knows they are the children of Israel, chosen by God. In exchange for giving them refuge, she asks only that they "save alive my father, and my mother, and my brethren, and my sisters, and all that they have, and deliver our lives from death." The spies promise to "deal kindly and truly with thee" when "the Lord hath given us the land."[2] They instruct Rahab to tie a scarlet thread in her window, and they promise that, when the Israelites arrive, "whosoever shall go out of the doors of thy house into the street, his blood shall be upon his head, and we will be guiltless: and whosoever shall be with thee in the house, his blood shall be on our head, if any hand be upon him." Rahab then lets the two men down from the wall "by a cord through the window."[3]

Later Joshua brings the Israelites across the Jordan and lays siege to Jericho.[4] God has instructed him to lead his army around the entire perimeter of the city once a day for six days; on the seventh day, "ye shall compass the city seven times, and the priests shall blow with the trumpets. And it shall come to pass, that when they make a long blast with the ram's horn, and when ye hear the sound of the trumpet, all the people shall shout with a great shout; and the wall of the city shall fall down flat, and the people shall ascend up every man straight before him."[5] The Israelites do as God has instructed. On the seventh day, at the sound of the trumpets, they shout, and the wall of the city falls down flat. The army enters Jericho, destroying "both man and woman, young and old, and ox, and sheep, and ass, with the edge of the sword." All except one household: Joshua instructs the two spies to find Rahab and her family and to bring them out of the town. The story ends with the words that Rahab "dwelleth in Israel even unto this day, because she hid the messengers, which Joshua sent to spy out Jericho."[6] Rahab's help to the Israelites earned her a place among the heroes of faith described in the New Testament, one of the only Gentiles so honored.[7]

The story of the wall of Jericho, and the decisive role played in its destruction by the woman who lives on top of it, is one of the best known in the Bible, and it is easy to see why. There is something infinitely satisfying about the image of an impregnable edifice brought down by sheer force of collective will, reason enough that "the walls came tumbling down" is perhaps, after "good fences make good neighbors," the most common catchphrase about walls. But it is also hard to ignore the violence of the story, which recounts the bloody conquest of one people by another. It is not only the wall that is destroyed, but nearly everything and everyone it protects, "both man and woman, young and old, and ox, and sheep, and ass, with the edge of the sword."[8] Joshua warns the survivors that, should anyone have the temerity to rebuild the wall, "he shall lay the foundation thereof in his firstborn, and in his youngest son shall he set up the gates of it."[9] Though the story is told from the point of view of the Israelites asserting what they see as a legitimate claim to the lands west of the Jordan, a modern reader's sympathies might equally lie with the people of Jericho whose city is besieged. For them the wall is not an obstacle to be surmounted—it is the one thing separating them from slaughter. The conquest of Jericho is thus not only a story of shared purpose and divine guidance: it is also a testament to the ancient link between walls and war.

It would be easy to leave it at this: the wall as instrument of violence and brute force. Yet the figure of Rahab suggests that this understanding is incomplete. Rahab is an unlikely hero, a person who helps men who have come to destroy the place she calls home. Her reasons for doing so are never stated explicitly; perhaps she simply wants to save those closest to her. But these reasons may also have something to do with her position on the social and political margins of the city, a position reflected in the place where she lives. A resident of the wall, Rahab belongs strictly to neither side of it, and it is precisely this position that enables her to mediate between the Israelite spies and the king of Jericho. Through the figure of Rahab, the wall is not a barrier but rather a conduit and medium of exchange, as she uses her window to send messages to the attacking force that all within her house should be spared.

But perhaps even more than the role of the wall as a conveyor of messages, the story of Jericho—the spies on the wall, the compassing of the city first six times, then seven on the seventh day—illustrates the closeness of walls to people's deepest beliefs, their most cherished rituals. Walls have always been these things too, even more than physical objects. This is not to say that walls have not also been associated with violence; the euphemisms the writer of the story uses hardly conceal this aspect. It is only to say that, like all the objects that people make, walls have never been any one thing. They have always been violence *and* nurture, separation *and* exchange, barbarism *and* mercy, destruction *and* creation—all at once.

The wall where Rahab made her home and to which the Israelites laid waste was not the first to surround Jericho, which was already an ancient city by Joshua's time. What is today called "Jericho" refers in fact to a number of different settlements, built and rebuilt many times in the same vicinity. The first evidence of human occupation at Jericho dates to about 10,000 BCE, when it was home to a group of sedentary hunter-gatherers; by 8000 BCE these first inhabitants had been replaced by a group of Neolithic farmers. During this period the town is estimated to have housed three thousand people and covered eight acres, enormous for the time. Many urbanists argue that Jericho is the oldest city, predating by several millennia the urban civilizations of Mesopotamia, Egypt, and India.[10]

Kathleen Kenyon, the archaeologist whose name is most closely associated with the excavations at Jericho, began by looking for traces of the walls described in the book of Joshua. She uncovered fortifications far older than those recounted in the Bible. The first "wall of Jericho" dates to 8000 to 7000 BCE, a monumental fortification consisting of a rock-cut ditch twenty-seven feet wide and five feet deep.[11] Behind this ditch was a solid stone rampart eighteen feet high with a massive stone

6 This wall did not tumble: 1950s excavation of the first wall of Jericho, built about 8000 BCE. Courtesy of the Council for British Research in the Levant.

watchtower.[12] The human effort required to build such a structure is almost inconceivable, particularly when the invention of the wheel lay millennia in the future.

The first wall of Jericho was a great feat of human ingenuity and purpose—no less so, indeed, than the Israelites' destruction of its distant descendant. What could have stirred the making of such an object? Like the construction described in the Old Testament, the earliest wall of Jericho almost certainly defended the settlement against human foes. In the eighth century BCE, Jericho would have been surrounded by similar but far smaller villages, many still with a mixed hunting and farming economy. Competition for land and resources was growing, largely as a result of the success of Neolithic agriculture. Yet the earliest settlement at Jericho precedes the development of clay pottery, considered a milestone because, without fired vessels, storing grain on any large scale is impossible. The first wall of Jericho would thus have created its own granary, a vessel for whatever surplus its inhabitants had been able to gather from the fields around it. This status as a repository of grain—and thus of life itself—was both the city's power and its chief vulnerability.

But saying that a city wall defended against an enemy in competition for land and resources is not the same as saying it existed only for that purpose. Like its biblical descendant, it is unlikely that the first wall of Jericho—whose construction demanded immense amounts of human labor and which would have endured over many lifetimes—would have been seen by people as performing a single function. Indeed, the very idea of "function" is suspect when applied to an ancient culture with mental constructs that were surely radically different from those of modern minds. The notion advanced in the 1930s by historian Arnold Toynbee—that human history and prehistory are made up of the environment's presenting "challenges" that humans respond to—is not sufficient for understanding why people would make such a monumental object as the first wall of Jericho.[13] Like the rest of the material culture of the place, the wall would have been bound up in the practices, perceptions, beliefs, institutions, and rituals of the people who made it. As archaeologists David Lewis-Williams and David Pearce have argued, there was no separation of "experience, belief and practice" at Jericho, or indeed in any other premodern society. Rather, the three formed an inseparable whole.[14] Walls were not "tools" designed to solve a problem, but complex and enduring parts of the symbolic system of the cultures where they arose. Surveying the long arc of human history and prehistory, then, it is necessary to ask not what walls did, but what they were.

Walls Were Nurture

On the Bay of Skaill, on the western coast of Orkney Mainland, the largest of the Orkney Islands north of Scotland, lies one of the richest archaeological sites in Europe. The Neolithic village of Skara Brae was built about 3000 BCE by people who had sailed north from the Mediterranean.[15] Buried in sand deposits for thousands of years, the village was revealed in the 1850s by a large storm and was partly excavated by archaeologist William Watt in the 1860s. It was not until the late 1920s, however, that Skara Brae was fully uncovered by Australian archaeologist Gordon Childe, and it is Childe's name that remains most closely associated with the sensational finds there.

Childe discovered that the village had consisted of eight family huts linked by a system of pathways. At its height, Skara Brae probably had about a hundred residents, who practiced the mixed hunter-gatherer and farming economy common to many Neolithic societies. Residents belonged to small family units that subsisted on the things they could harvest from land or sea, as well as the animals they had domesticated and the grains they had brought up the coast of western Europe. The dwellings were made of sea-quarried stone, their roofs of whale skin stretched over beams of whale ribs (there was no wood on the island). Childe also found much of the contents of dwellings intact, since these items too were made of stone: stone beds, stone shelves, and stone containers for storing grain. Skara Brae seemed a complete village, with even a rudimentary system of reticulated sewers.[16]

Childe and subsequent archaeologists noted one striking absence at Skara Brae: there were no discernible shrines or other religious structures. Even altarlike articles of furniture in some dwellings, they believed, were used not for numinous purposes but for showing off to visitors the household's prized objects. So where were the sacred precincts, the places for communicating with the spirits of ancestors? Where were the temples, the shrines, the altars at which gods were revered and propitiated? They were embedded, it seems, in the walls.

Though the residents of Skara Brae had ample materials to build their dwellings aboveground, the village was partially subterranean, with only the roofs of structures visible on the surface. But the village was not buried because it had been excavated. Rather, it appears that villagers built the walls of their stone dwellings aboveground, then mounded what archaeologists call "artifactual and ecofactual" wastes around them. In other words, the residents of Skara Brae created and reinforced their enclosures with their own household refuse. Since the midden was essentially a compost pile, it created heat, warming the dwellings within, and protected the huts against the storms and winds of the North Sea. It also improved the structural soundness of the huts; later excavations at Skara

7 Wall as nursery: family dwelling, Skara Brae, Orkney, Scotland, about 3000 BCE. Photograph by Thomas Oles.

Brae revealed that midden materials, some containing bone, were also used as an insulating core and mortar for the walls of dwellings.[17]

A modern consciousness finds it easy to discern these practical benefits. But such easily identifiable functions do not in any way preclude the possibility that there were other meanings to midden construction at Skara Brae. Archaeologist Ian Simpson has noted that using the material residue of the past in building likely created for the inhabitants of the village a "relationship with previous generations and [a] sense of place."[18] If this is so, then we must abandon the very notion of the middens as "waste" in the modern sense. The middens were part of dwellings, and thus inextricably related to the beliefs of the people who lived within them, even if the precise contours of that relationship are now lost forever.

Another striking feature of Skara Brae were the beds, longish rectangles made of three large stone slabs set on edge, with the wall of the dwelling as the fourth side. The Neolithic people who inhabited Skara Brae would have been small, but these beds were too short even for them; it is likely that they slept sitting up. But the most interesting feature of the beds was buried in the walls beside them. Here, in several cases, Childe discovered crypts filled with human bones, arranged in exactly the same sitting position as the person sleeping next to them. The wall was quite literally made not just of the *objects* one's ancestors might have used, but also of one's ancestors. The dead were likely buried after *excarnation*, a method in which corpses are exposed to scavengers and the elements before being buried. Once the bones had been picked clean and bleached, they were placed in the designated enclosures inside the wall, sleeping beside their living relatives. The dwellings of Skara Brae were thus both houses and tombs, their walls not so much mediators between the living and the dead as the bond between them.

Skara Brae appears to have lived at peace with its scattered neighbors. There appears to have been no contest for resources: no human bones bear the marks of violent death, and the village had no significant defenses of any kind. This did not mean, however, that there was no concern with enclosure among the people who inhabited the village. Many archaeologists have noted the symbolic role in Neolithic cultures throughout the world of containers of all kinds, from pots, jugs, and granaries to the human womb, represented in the bloated figure of the earth goddess.[19] Although the production of such vessels blossomed with the development of fired clay pottery, even Neolithic villages like Skara Brae offer evidence of this preoccupation with enclosure, girdling, containment. In addition to vessels for storing grain and other foods, the walls of each hut contained a wide array of recesses and niches whose precise use and meaning remain unclear to archaeologists. The beds of the dwellings, like the tombs that lay beside them in the walls, were merely another kind of niche.

No less than the defensive wall of Jericho, the non-defensive walls of Skara Brae would have played a critical role in ensuring a continuous food supply. This characteristic would likely have made them potent symbols of security and predictability for the residents of the village, or indeed of any Neolithic settlement.[20] Where a village was enclosed by an earthen bank or midden, human life could flourish in unprecedented ways: children could play without being threatened by animals, grain could be stored safely from year to year, and livestock could be domesticated, dependent on their human masters for the protection the village offered. The fermenting middens around the walls of dwellings would have made the temperature milder inside the village than outside. It is hard to imagine that these "functions" of walls would not have made them psychic anchors in the cosmos. The walls of Skara Brae made the village a single vessel, bearing its hundred or so passengers, living and dead, into a future that was ever so slightly less uncertain because of

8 Marginal ecology: ancient hedge and stile, Cornwall, England, about 1500 BCE. Courtesy of Derek Harper.

their existence.[21] They embodied the possibility of life itself. They described not a redoubt, but a nursery.

At the other end of the British Isles, near Land's End in Cornwall, spreads an extraordinary landscape. A network of massive earthen banks—seven feet high and over fifteen feet wide; topped with boulders, shrubs, and even trees—divides the green slopes into small irregular pastures. The banks, which contain no gates and can be crossed only at stiles cleared of plants and debris, follow the rolling topography, disused banks extending like fingers onto the heath above.

These banks are the only remaining artifact of an agricultural society that flourished in this part of the British Isles in the early Bronze Age. Using little more than their hands, the inhabitants of this culture pushed the granite boulders that had rolled down from the moor into lines; when they encountered a boulder or "grounder" too large to move, they simply altered the course of the bank.[22] When they had set the largest boulders, they piled smaller stones on top of them, filling remaining crevices with loose soil. Over the centuries, farmers planted the banks to create nearly impenetrable barriers. Once in place, these boundaries were impossible to alter. It is not known whether the resulting enclosures were proprietary, but it is certain that they held livestock because they still serve this purpose. This makes them, at over three millennia, among the world's oldest continuously used objects.

These walls were an accomplishment no less great than the walls of Jericho or Skara Brae. They would have demanded immense collective effort in any modern society, to say nothing of an ancient one; that they were built across an entire landscape suggests how much they benefited the people who made them. The earthen banks protected flocks from predators and weather, and when flocks flourished, so did the people who tended them. These walls thus embodied the possibility of human life and increase.

Such associations are likely to have left a deep imprint on human minds. It is of course impossible to know how a Bronze Age farmer would have understood the embankment that protected his livestock, fed his household, and nurtured his community, but material evidence suggests that the banks were important sites in this culture. Archaeologists have discovered that residents buried many household objects in the banks, just as residents of Skara Brae had used the walls of their dwellings to inter their ancestors centuries before, practices that suggest the intimate link between walls, belonging, and belief.[23]

The earthen banks of Cornwall are but one example of a pattern that spans places, times, and cultures. Making enclosures to protect livestock, grow crops, or define territory is one of the oldest ways people have shaped the earth. For millennia, humans have created such enclosures from the materials the land offers, pushing soil into berms, adding stone found nearby for support, and planting the resulting construction with shrubs and trees. The original English term for such practices is "hedging," a word that derives from the Saxon *gehægen*, "enclosure."[24] In medieval England, when one spoke of a "fence" one was most frequently talking about what people today would call a hedge.

The Cornish banks suggest how greatly hedges in this original sense would have differed from the neat plantings around yards and gardens today. As late as the early nineteenth century, agricultural observer Charles Vancouver could give this description of a recently built hedge he had seen in the county of Devon:

> The fences were raised upon a base seven feet wide, with a ditch of three feet on each side, and which, including the foot for the sods or facing to rest upon, occupied about thirteen feet width of ground. The mound was raised six feet high from its base; the

sides faced with turf and left nearly five feet wide on the top: these were planted with two rows, consisting of oak, ash, beech, alder, hazel and hawthorn.[25]

A construction like this would have been a living, growing entity that, at maturity, presented a nearly impenetrable barrier to both livestock and people. Yet containment of animals only begins to suggest the ways in which hedges have nurtured rural economies throughout history. Not only did hedges keep animals from wandering across cultivated fields or into yards; they reduced wind speed, retained soil moisture, and prevented erosion, all of which improved crop yields.[26] In medieval England, blackberries, elderberries, sloes, rose hips, and beechnuts were gathered from hedgerows.[27] "Leafy hay" harvested from hedges was used as fodder for livestock, and the common hedge plant stinging nettle was used for food, medicine, and fiber.[28] Hedges also provided fuel and building material, and many were coppiced in the same way as neighboring woodland.[29] Oaks, ashes, and elms left to grow to maturity in the hedgerow, called "standards," were periodically felled for timber.[30] Such uses were enshrined in the common law right of "hedgebote," which persisted in some rural communities in England until the nineteenth century. Hedges were far more than markers of property or barriers to livestock. They were an inalienable part of the human economy as a whole, nurturing the social landscape as much as the physical one.

From Skara Brae to medieval England, walls created the conditions and provided the materials for human survival, protecting people from both predators and the elements.[31] It is easy to imagine how this might have made them potent symbols for the people who built them. Yet it is a small step from control over the natural environment to control over people and land. A line thus leads from the banks of Neolithic Europe to the earliest civilizations—and the kings who ruled over them.

Walls Were Message

The Epic of Gilgamesh tells the story of the semi-mythical king who ruled at Uruk, in southern Mesopotamia, during the first half of the third millennium BCE. The only extant version was written in the Akkadian language in cuneiform script, on twelve tablets found in the nineteenth century in the ruins of the library of the Assyrian king Ashurbanipal at Nineveh, in modern Iraq. The epic recounts how the god Anu, wanting to curb the overweening pride of Gilgamesh, creates the wild man Enkidu. Gilgamesh and Enkidu fight, and Enkidu is vanquished. After this defeat, however, Enkidu becomes the king's companion and servant. The poem recounts the adventures of the two in the "Land of the Cedars," probably modern Lebanon, and attempts by Ishtar, the goddess of love, to seduce Gilgamesh when the two men return to Uruk. Enkidu falls ill, and on his deathbed he relates to Gilgamesh a dream of the underworld that awaits them both. Gilgamesh has been given kingship but not immorality, and the end of the epic tells how "all men of flesh and blood lift up the lament" on hearing of his death.

The Gilgamesh story is a great literary adventure that is still read today by adults and children alike. But it is also an invaluable source of information about Sumer, the first urban civilization, which arose in the valley of the Tigris and Euphrates about 3300 BCE. The epic is rich with detail on the culture, values, and myths of the city-states that made up Sumerian civilization, and it sheds considerable light on other early urban cultures as well. Indeed, it begins not with the figure of Gilgamesh himself, but with the city associated with him and its great circuit of walls:

> In Uruk he built walls, a great rampart, and the temple of blessed Eanna for the god of the firmament Anu, and for Ishtar the goddess of love. Look at it still today: the outer wall where the cornice runs, it shines with the brilliance of copper; and the inner wall, it has no equal. Touch the threshold, it is ancient. Approach Eanna the dwelling of Ishtar, our lady of love and war, the like of which no latter-day king, no man alive can equal. Climb upon the wall of Uruk; walk along it, I say; regard the foundation terrace and examine the masonry: is it not burnt brick and good? The seven sages laid the foundations.[32]

Besides ranking among the great literary openings of any age, this is the fullest depiction of a city wall from an ancient civilization. For its writer (almost certainly a priest, since only priests and kings were literate), the wall of Uruk is inextricably linked with the extraordinary figure, "two-thirds god and one-third man," who oversaw its construction. The wall is an expression of that divinity, reinforced by references to the other members of the Sumerian pantheon: Anu, Ishtar, and the "seven sages" who laid the foundations. Only gods, it seemed, could have sanctioned and wrought such a structure.

But these words not only describe how the wall expressed the extraordinary power of a single god-king; they also show how the wall of this city, indeed any city, would have symbolized, for all those who encountered it, the technical, administrative, economic, and military power of urban civilization as a whole. The epithet "strong-walled Uruk" recurs throughout the Gilgamesh epic, supporting Max Weber's contention that walls and cities are originally synonymous.[33] Walls produced the compression necessary to ensure the essential collisions of urban culture, which could never have arisen in widely dispersed villages such as Skara Brae or those around Jericho. The tone of civic pride, even awe, in the opening passage is a function of that association. For the author of the Gilgamesh epic, the wall of Uruk embodies the splendor of a society that was in many ways more advanced than any that had come before it. The wall's foundation terrace, parapet, and cornice, its burnt brick shining "with the brilliance of copper"—these were not merely material descriptions but symbols of Sumerian civilization, which had reached unprecedented heights of architecture and metallurgy.[34] The wall of Uruk was a message to the world that has carried across the ages.

Geographer Robert Sack has suggested that territoriality in human culture has several essential components. The first is classification by area, or association of some part of the earth's surface with particular things, people, or events. Another is enforcement, the ability to control access to this area by either physical means or the threat of them. Both these components depend, however, on the boundary. Boundaries are arguably the central device of territoriality, because they are a form of symbolic communication that combines "direction in space and a statement about possession or exclusion."[35] Every boundary thus assumes some audience, an interlocutor to whom its meaning is addressed within a context of understood rules and conventions. In short, every boundary ultimately is a message.

This notion is consistent with the tendency, widespread in Mesopotamia, to build city walls so massive and imposing that they exceeded any military need. The wall of Uruk was as much a public expression of power and wealth as a defense against enemies. "The heavy walls of hard-baked clay or solid stone," Lewis Mumford noted, "would give to the ephemeral offices of state the assurance of stability and security, of unrelenting power and unshakable authority."[36] This symbolic function of the wall appears to be confirmed by early military practices in Sumeria. In the period when Gilgamesh reigned at Uruk, walled cities were seldom attacked directly, and sieges were rare before the arrival of the Assyrians in the eighth century BCE. Battles were fought instead in the lands between cities, with rival armies sallying out of their respective enclosures.[37]

In other cultures the reverse has been the case, with trompe l'oeil being used to impress, and ward off, potential foes. Cadbury Castle, an Iron Age fortress not far from the earthen banks of Cornwall, had a defensive perimeter of over one kilometer, a stockade set atop a semi-circular rampart faced with limestone, which would have appeared solid to any attacker approaching from below. But archaeological investigations have revealed that this defensive perimeter was very weak, and that stones that looked insurmountable from a distance were not even keyed into the mortar between them.[38] Even the wall that the Roman emperor Hadrian built across England in the second century CE was as much an expression of Roman imperial power as a physical barrier to the Pictish people north of it. Far from blocking contact, Hadrian's Wall assumed it, for only then would its primary message—that breaching the wall was not worth the effort—be received.[39] In this it was similar to the other great territorial wall of history, the so-called Great, or Long, Wall of China, constructed during the Ming dynasty in the sixteenth century CE. According to historian Arthur Waldron, while it may have looked impressive from afar, the Great Wall was primarily a massive symbol of dynastic power and nascent Chinese identity, "useless militarily even when it was first built."[40]

But walls do not bear only the messages their builders intended. Set into the first city wall of the ancient Mesopotamian city of Ur were bricks stamped with the initials of the Persian king Cyrus the Great, who conquered the city in the sixth century BCE. Each brick bore the message, "The great gods have delivered all the lands into my hands."[41] The Gilgamesh epic, too, suggests that such an imprimatur had been common practice for millennia. Before he sets out to battle the monster Humbaba in the "cedar forest," Gilgamesh laments to Enkidu, "I have not established my name stamped on bricks as my destiny decreed; therefore I will go to the country where the cedar is felled."[42] City walls were not simply barriers against foes: they were signboards for announcing conquest and control.

And yet, as grand a statement as a city wall was, the rise of urban civilization depended on another kind of wall altogether. All the kings and priests, scribes and artists, masons and smiths—all the inventions of geometry, mathematics, and writing that made cities cities—depended on the wealth and surplus created in the fields and pastures that spread beneath the city wall. Alexander the Great noted to the architect Dinocrates in the fourth century BCE that "just as a baby cannot nourish itself and grow without its nurse's milk, neither can a city without fields and produce flowing into its walls."[43] The true source of urban culture was not the wall of the city, but the walls, hedges, and ditches marking the lands around it.

The first three urban civilizations—Mesopotamia, Egypt, and the Harappan culture of the Indus Valley—arose in fertile alluvial plains. Even with their flourishing non-agricultural specializations, agriculture still dominated these societies. Most urban residents in Mesopotamia, for example, still raised crops outside the city wall, and the fortunes of Ur, Uruk, and Babylon rose and fell with the yearly harvest.[44]

Agriculture in each of these early urban civilizations depended on elaborate systems of irrigation, not only to deliver a regular supply of water to fields but also (since every ditch implies an earthen bank alongside it) to enclose livestock and delineate parcels of land. A cadastral map dated roughly 1500 BCE for the fields outside the city of Nippur, in the southeastern part of modern-day Iraq, suggests how extensive such canals must have been. It shows a bend in a river and fields the boundaries of which are marked by double incisions representing ditches.[45] These irrigation systems were fundamentally unstable, frequently damaged or obliterated by floods of the Tigris, Nile, and Indus, and boundaries could be counted on to shift from year to year. A more persistent boundary marker was needed than merely the ditch itself. These early civilizations thus all used boundary stones embedded deep in the earth to mark edges of fields and minimize conflict among the people who tended them. Particularly in Egypt, with its regular yearly flood, there developed a sophisticated system of yearly land survey—according to Herodotus, the basis of modern geometry—to restore boundary stones and levy taxes.[46]

Tampering with a boundary was a serious infraction in these early civilizations. Boundary stones often bore minatory messages addressed to anyone who might consider disturbing them, with long descriptions of the elaborate punishments awaiting that person in the afterlife. For example, a Babylonian boundary stone, or *kuduru*, from the reign of Meli-Shipak in the late second millennium BCE reads:

> Whenever in future times . . . one shall rise up . . . and . . . shall bring an action, or make a claim or cause a claim to be made, or shall send another and cause him to take or lay claim to, or seize it or shall say "This field was not granted" or the boundary stone of that field, through any wickedness shall cause a fool or a deaf man or one who does not

9 Legal boundary as divine admonition: *kuduru* documenting title to an estate, Babylon, Mesopotamia, about 950 BCE. Courtesy of Erich Lessing/Art Resource, New York.

understand, to destroy or shall change it, or break it up. . . . May all the gods, whose names are mentioned on this boundary stone destroy his name, and may they bring him to naught.[47]

The cuneiform script in which this warning was conveyed, used only by kings and priests, would have imbued the stone with even greater authority. Boundary stones were thus more than simple markers of territory: they were messages backed by divine writ.

But in no civilization did boundary stones play such an important role as in Rome. In the first *Georgic*, Virgil looked back "before Jove's time" when "no farmer plowed the earth" and "it was forbidden to mark out field from field, / Setting out limits, one from another."[48] This world without limits had long since vanished when Virgil wrote these lines in the final years of the Republic, field boundaries having been clearly marked for centuries with stones called *termini*. As Rome grew, conquering new lands and peoples, termini became the essential record of centuriation, the surveying and subdividing of territory at which the Romans excelled. Land-dispute arbitrations from the first century CE onward suggest just how littered with termini the Roman landscape had become. One decision, setting a boundary near modern-day Genoa, reads: "There is a boundary stone at the confluence of the Edus and the Procoreba. From the R. Edus to the foot of Mt. Lemurinus, where there is a boundary stone. Then straight up the ridge of Mt. Lemurinus; there is a boundary stone on Mt. Procavus"—and so on.[49]

Like their counterparts in Mesopotamia, Roman termini bore inscriptions investing them with state authority, and they often referred explicitly to the emperor reigning at the time the stone was set. Such inscriptions served as a durable public record about the constitution of colonies; termini dating from the second triumvirate at Capua, for example, fix the location of the sacred "first furrow" around that settlement with the words: "By order of Octavian, on the line plowed."[50] Moving an object with such an imprimatur would have subjected any person to severe penalties.

Termini not only bore messages of imperial power; they also gave practical information about the topographical and hydrological features of particular places. One treatise titled *De terminibus* in the *Corpus agrimensorum*, the principal Roman treatise on surveying, recommended that boundary stones be set not just along the edges of fields, but wherever surveyor's lines crossed natural boundaries such as springs, streams, wells, groups of trees, ravines, and rises. The treatise provided guidance on how to read these messages: "If a boundary stone is hewn square and has a dot on it, it indicates a spring. But if it has a hollow on top, it indicates a well at the boundary."[51] Boundary stones also provided information about the social and political landscape. The Roman surveyor Hyginus Gromaticus described having to saw up stone altars "recording on the side facing a colony the boundary of that colony, on the other side the name of the neighboring townspeople."[52] Some termini bore the inscription FPV, meaning "farm of earlier occupant," while others conveyed elaborate restrictions on use of parcels ("The lower road is private property belonging to Titus Umbrenius son of Gaius. Entry on foot by permission; no cattle or carts").[53]

Termini were also deities. Terminus was one of the key gods of the pantheon the Romans had inherited from the Etruscans, whose deities were not anthropomorphic but were mystical presences inhering in objects, or what the Romans called *numina*.[54] Termini were thus stones, boundaries, and gods all at once. This numinous aspect appears also to have been present in Babylonian culture, where sacrificial remains were buried beneath *kuduru*, a "chthonic implication which is echoed by the fact that the violator of boundaries is 'damned' to the infernal gods."[55] Moving or damaging a boundary stone in both cultures subjected the offender not only to legal penalties, but also to the wrath of the deity who oversaw it.[56]

This identity of boundaries and gods suggests how walls have always done more than simply convey messages about territory and control. They have also been channels for religious experience. Walls of all kinds—from fields to cities to dwellings—were always bound up with basic structures of human belief.

Walls Were Belief

The remains of the ancient city of Çatalhöyük lie on an old branch of the Çarsama River, three thousand feet above sea level in the shadow of the Taurus Mountains, in south-central Turkey.[57] The settlement is thought to date from approximately 6500 BCE, when farmers who had begun to cultivate barley and emmer in the surrounding hills brought agriculture down to the Konya Plain, where they also hunted wild aurochs, pigs, and deer.[58] Çatalhöyük profited from the large obsidian deposits in the mountains, and it appears to have been an important node on the trade route linking southern Anatolia to the eastern Mediterranean as early as the eighth century BCE. (Tools made from obsidian of Anatolian origin were also discovered at Jericho, the only settlement that can contend with Çatalhöyük for status as the world's oldest city.)[59] This combination of agriculture and trade made Çatalhöyük an extraordinarily large and prosperous settlement for its time, covering forty acres and housing up to six thousand inhabitants.[60]

Çatalhöyük ranks alongside Ur as one of the most important archaeological discoveries of the twentieth century. James Mellaart, the archaeologist who unearthed the site between 1961 and 1964, discovered twelve layers of Neolithic settlement extending fifty feet below the surface, corresponding to nearly one thousand years of continuous occupation before the city was abandoned about 5600 BCE.[61] The society that built Çatalhöyük was conservative in shaping its environment. Though construction methods were refined over the centuries, each layer of settlement was remarkably similar, consisting of adjacent single-room houses of timber and mud brick stepping down the sloped site in orderly ranks. Each structure appears to have housed a family of four to eight members. Houses were entered by ladder from a single hole in the roof, and the only way to move across the city was by climbing from roof to roof. The sole open spaces were courtyards where older buildings had collapsed and where residents dumped ash and household refuse; the interiors of the houses, by contrast, were orderly and spotless.

Like Childe at Skara Brae, Mellaart found no evidence of a designated shrine or sacred precinct at Çatalhöyük. Instead, he discovered religious imagery built into the structure of residences. Approximately forty houses contained elaborate wall paintings depicting domesticated and wild animals, particularly aurochs, the predecessor of modern cattle. Even more striking, many of these walls bore ceremonial aurochs heads, or *bucrania,* that served as the focus of domestic altars. Mellaart speculated that these images and sculpted forms were devoted to the cult of a male deity, and he suggested that the walls of Çatalhöyük represented a hitherto missing link between Upper Paleolithic and Neolithic cultures.[62] A particularly important part of the rituals surrounding these wall paintings appears to have been a process of replastering and "laying on hands" in which residents coated their hands with red paint and pressed them to the wall. The palm prints that remained almost certainly had religious as well as decorative significance. People placed their hands on the walls because those walls were "a membrane between them and the spirit realm."[63] The numinous power of these imprints is further suggested by their having been ritually defaced when a structure was abandoned or demolished.

Perhaps the most striking single wall painting at Çatalhöyük, however, was a proto-landscape. Along one wall of a private room, Mellaart discovered a large image of the entire city and its surroundings, drawn with striking accuracy and to nearly correct scale. The twin peaks at the upper edge of the image are clearly recognizable as Hasan Dağ, the volcano whose obsidian deposits were the source of the city's wealth and probably the reason for its existence. This wall painting is believed to be the first map of a human settlement and landscape.[64]

The walls of Çatalhöyük were not simply shelter, defense, or handy

10 Membrane of the spirit world: shrine with aurochs heads on walls, Çatalhöyük, Anatolia, about 6150 BCE. Courtesy of Ida Pedersen.

surfaces where totemic animals or rituals could be painted. They were places of numinous emanation, membranes between divine and human, dead and living, that did not separate these realms but bound them ever more tightly together. As the wall map of the settlement suggests, walls were nothing less than models of, and guides to, the cosmos. This essential aspect of walls would not end at Çatalhöyük but would come to mark every civilization that arose in its wake.

At the opposite end of Anatolia from Çatalhöyük, in the late fifth century BCE, the Spartan mercenary Lysander paid a visit to the Persian prince Cyrus the Younger at his imperial residence at Sardis, capital of Lydia. The two men were friends and allies, united in opposition to Athens during the Peloponnesian War. As later recorded by Xenophon in the *Economist*, Cyrus received Lysander in an extraordinary environment the likes of which the latter had never seen:

Lysander, it seems, had gone with presents sent by the Allies to Cyrus, who entertained him, and amongst other marks of courtesy showed him his "paradise" at Sardis. Lysander was astonished at the beauty of the trees within, all planted at equal intervals, the long straight rows of waving branches, the perfect regularity, the rectangular symmetry of the whole, and the many sweet scents which hung about them as they paced the park. In admiration he exclaimed to Cyrus: "All this beauty is marvelous enough, but what astonishes me still more is the talent of the artificer who mapped out and arranged for you the several parts of this fair scene." Cyrus was pleased by the remark, and said: "Know then, Lysander, it is I who measured and arranged it all."[65]

The old Persian word that Lysander would have heard Cyrus use to refer to this place was a combination of two roots: *para*, "around," and *daeza*, "wall." The word was taken by the Greeks and turned into *paradeisos*, later used in the Greek translations of the Old and New Testaments. Like the modern English "garden," from the Saxon *gærd*, or "wattle fence," "paradise" is by its very definition a walled place, a precinct set apart from the world.[66] The benefit of such an enclosure principally had to do with its capacity to sustain life in a harsh and unforgiving environment. Every paradise first arose around an oasis. By casting shadow and blocking wind, its mud-brick walls created a cooler, moister climate around the water source, giving rise to a refuge where fruit trees and edible plants could be cultivated. Both Cyrus and his brother Artaxerxes II used their paradises to house exotic tree species and menageries of animals brought back from distant campaigns.[67] Such menageries were more than just curiosities that would never have survived outside the garden: they were attempts to collect all of God's creation in one place.

A paradise was accessible only to the few. It was this *separateness* of the precinct inside the wall that made gardens key loci of political power all over the ancient world. Paradise was not only a place of spiritual repose for the elite and a container for curiosities assembled from distant travels, but also a stronghold from which kings and princes might administer their own realms and plan the conquest of others; the occasion of the garden meeting between Lysander and Cyrus, after all, was to plot their final campaign against Athens. Paradise, ultimately, was a seat of power—power that was inextricable from its enclosure of the source of life itself.

A paradise wall, shimmering in the heat on the distant horizon, would thus have symbolized both physical sustenance and political power. But it is too simple to leave it at that. In their original form, gardens were more than just protected oases. They were what anthropologist of religion Mircea Eliade famously called *axes mundi*, places where communication was possible between the earthly and divine realms.[68] Gardens are thus closely linked to shrines, temples, and burial grounds: all are places that are simultaneously *of* the human world and *distinct* from it. A garden wall was not simply a fortification. It was the symbol

11 Scarcity and succor: walled oasis and caravansary, Persia, about 1600 BCE. Courtesy of Harpur Garden Images. Photograph by Jerry Harpur.

of the very possibility of communication with the spirit world, the embodiment of the heavenly order that the garden as a whole represented.

The city wall of Uruk is not the only wall mentioned in the opening of the Gilgamesh epic. The author also asks his listeners to contemplate the "temple of blessed Eanna for the god of the firmament Anu, and for Ishtar the goddess of love." Eanna was what the Greeks called a *temenos*, or fortified religious citadel at the core of the city. Reconstructions by archaeologists show a precinct protected by high corrugated walls of sun-dried brick with the temple building at its center, characteristics common to such districts throughout the cities of the Mesopotamian world. From the time of Nebuchadnezzar, the *temenos* of Ur was enclosed by walls scarcely less massive than those of the city itself. As described by Leonard Woolley, the wall was "33 feet wide and probably about 30 feet high, the face of it was decorated with the double vertical grooves which were traditional for temple walls, and it was pierced by gateways of which six have been discovered; the main gate, with a high gate-tower set back in a deep recess, led immediately to the entrance of Nannar's chief temple."[69] At the corner of this sacred precinct rose the ziggurat, a pyramidal altar with walls hundreds of feet thick at

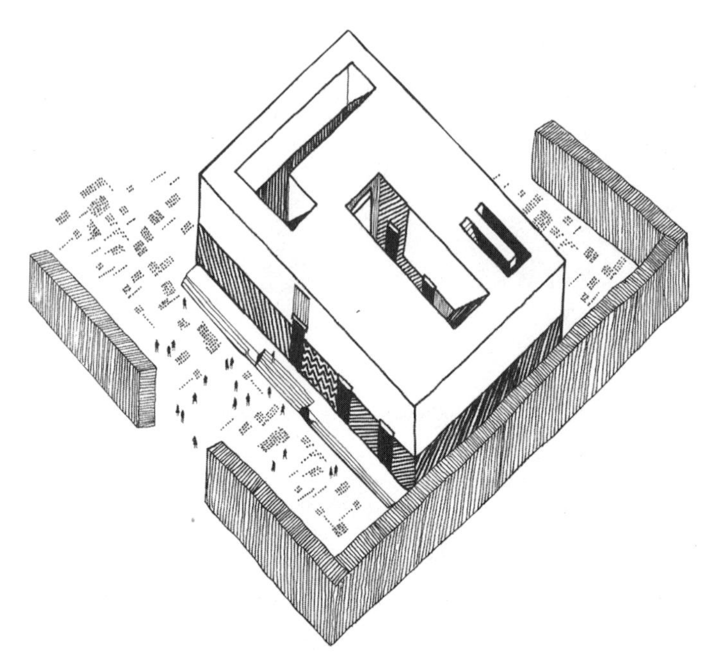

12 Pride of Gilgamesh: Eanna temple precinct, Uruk, Sumeria, about 3500 BCE. Courtesy of Ida Pedersen.

the base.[70] Like the city wall, such sacred edifices could only have been constructed and maintained over many years and at substantial cost in labor and life.

A modern consciousness may find it difficult to discern the need for temple walls as solid as those described in the Gilgamesh epic and discovered throughout Mesopotamia. But this misses the point. Neither the city wall nor the inner temple wall was merely defensive; their function (if indeed the concept of "function" can even be applied to premodern societies) was equally, if not primarily, to inspire reverence. If the city wall symbolized the political power of the king, then the temple wall embodied the spiritual power of the gods, with whom the king was allied and consanguineous. This semiotic identity of city wall and temple wall was noted by Eliade: "Long before they were military erections, [city walls] were a magic defense, for they marked out from the midst of a 'chaotic' space, peopled with demons and phantoms . . . an enclosure, a place that was organized, made cosmic, in other words, provided with a 'center.'"[71]

This notion is borne out by the frequent connection throughout history between city and temple walls and the deeds of potent mythic beings. The Greeks believed the immense curtain wall of Mycenae had been assembled by the Cyclops, thus giving the name "Cyclopean masonry" to the dry-stack construction the Mycenaeans excelled at. The Incas, New World masters of the same technique, are often said to have designed the indestructible wall of their capital, Cuzco, in the form of a puma, one of the totemic animals of Inca culture. The wall of the Chinese city of Ch'uan-chou Fu was built to resemble the carp, an "auspicious creature," "in a geomantic effort to secure prosperity and good fortune for the city."[72] And the Western Wall, the only remnant of the Second Temple of Jerusalem, has been the site of Jewish devotions since the early Byzantine period. To this day rabbinic belief holds that "the divine Presence never departs from the Western Wall."[73]

The walls of the first cities were thus not simply useful and sometimes indispensable constructions, but sites of numinous awe. This was likely as true for visitors as for the city dwellers. To an approaching trader (or army) descrying it on the distant horizon, an encircling wall such as that of Ur or Uruk would have resembled nothing so much as the handiwork of the gods. Similarly, for those inside, the wall would have been far more than an instrument for defense against human foes. It would have been the earthly manifestation of a divine realm. As at the proto-city Çatalhöyük, city walls mediated between that realm and everyday, waking life. If the city was, as Lewis Mumford wrote, a "man-made replica of the universe," then the wall was its earthly frame.[74]

The importance of this numinous aspect of city walls is suggested by one of the earliest maps in existence, a Babylonian map of the world dating from approximately 600 BCE. Now housed in the British Museum, this map shows a circular world bounded by an all-encompassing "Salt Sea," clear reference to the Persian Gulf. Two vertical lines bisecting the circle represent the Euphrates, flowing from mountains at the top to the swamps of lower modern-day Iraq at the bottom.[75] From the perimeter of the band radiate three of what probably were originally eight triangles, each depicting a region populated by legendary beasts described in incised text above the figure. Finally, straddling the river just above the center of the circle is a rectangular figure representing the enclosing enceinte of Babylon. The city wall, in short, stood at the center of the Babylonian cosmos.

The connection of walls and belief is impossible to separate from the very real things walls did to ensure human increase. Whether by modifying climate, storing grain, enclosing stock, or protecting from predators, walls made survival in the face of enormous environmental odds at least slightly more likely. The spiritual aspects of walls are inextricable

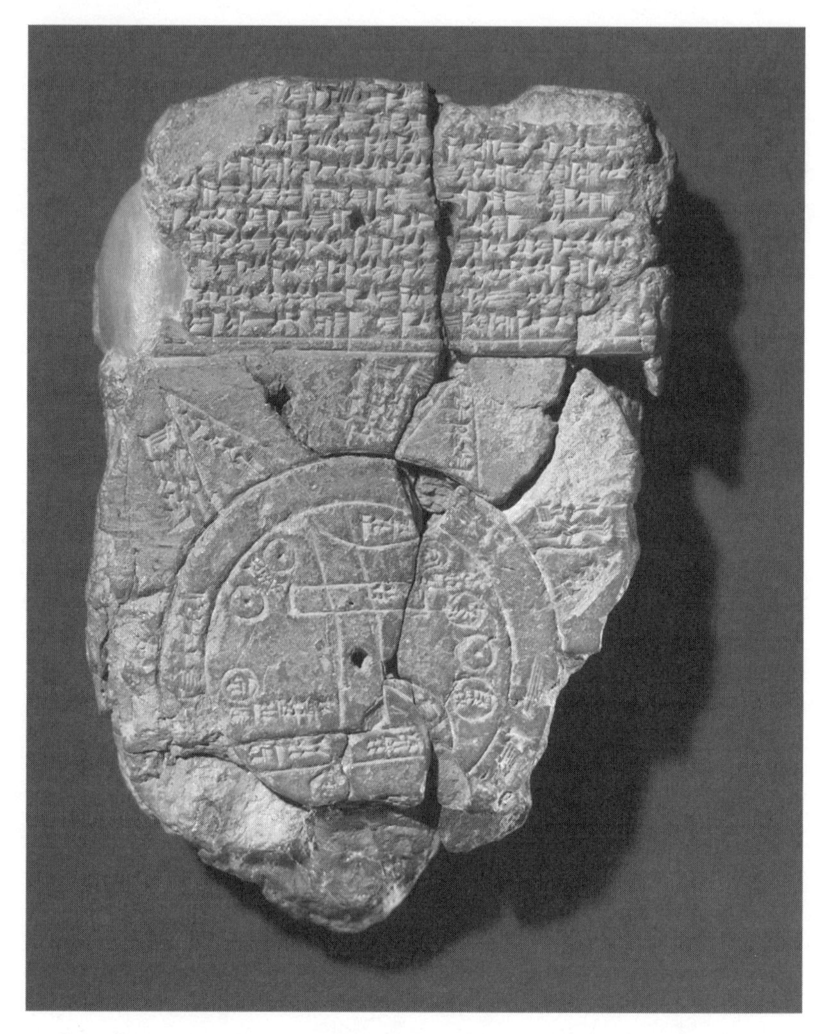

13 City wall as cosmic cynosure: world map, Babylon, Mesopotamia, about 600 BCE. Courtesy of the British Museum.

from these practical ones; indeed, it would not have been possible for a premodern person to draw such a distinction. From Skara Brae to Uruk to Babylon, walls were part of people's spiritual apparatus, not because there was some point at which, after discovering their many uses, people chose to invest them with this power, but rather because they were integral to the material and psychic culture of the societies that built them. In short, walls have always been an inalienable part of human dwelling.

Walls Were Dwelling

In the 1930s and 1940s, anthropologist Claude Lévi-Strauss traveled widely throughout the Amazon basin, trips he described vividly, and sometimes poignantly, in classics such as *The Elementary Structures of Kinship*, *Structural Anthropology*, and *Tristes Tropiques*. The societies Lévi-Strauss encountered on those trips were among a vanishing number in the world whose culture could still be described, at the time, as Neolithic. Among these societies were the Bororo, a people who lived in a series of villages along the Rio Vermelho and its tributaries.[76] The Bororo already had a long history of contact with Europeans, particularly the Spanish Catholic missionaries who attempted to convert them beginning in the eighteenth century, but it was Lévi-Strauss who introduced their culture to a world audience.

Lévi-Strauss devoted considerable attention to understanding the relation between the social structure of the Bororo and the physical structure of their villages. Every village, he found, was composed, exactly and invariably, of twenty-six huts arranged in a loose circular formation around a central "men's house," a "home for bachelors and a meeting place for the married men" that was strictly forbidden to the women of the village. The men's house was fronted by a large open space of beaten earth used for dances and ritual celebrations. Multiple radial paths led away from this central space to the twenty-six family dwellings. These huts were "matrilocal," or identified specifically with mothers and children; the opposition between the center and the periphery of the village was thus also an opposition between men and women.[77] The boundary of every Bororo village was thus made from the dwellings of villagers. Not only was this boundary associated with women, Lévi-Strauss found, but its form and organization embodied the entire matrilineal system of kinship relations among villagers. Every Bororo village was divided along an east-west axis into two "moieties," each consisting of four family units. This division ran perpendicular to the watercourse on which the village lay, with the resulting halves of the village invariably referred to as "upstream" and "downstream." This meant that at any given point along its circumference, the boundary of the village was marked with respect to gender, the clan of its inhabitants, and its position in the surrounding topography.

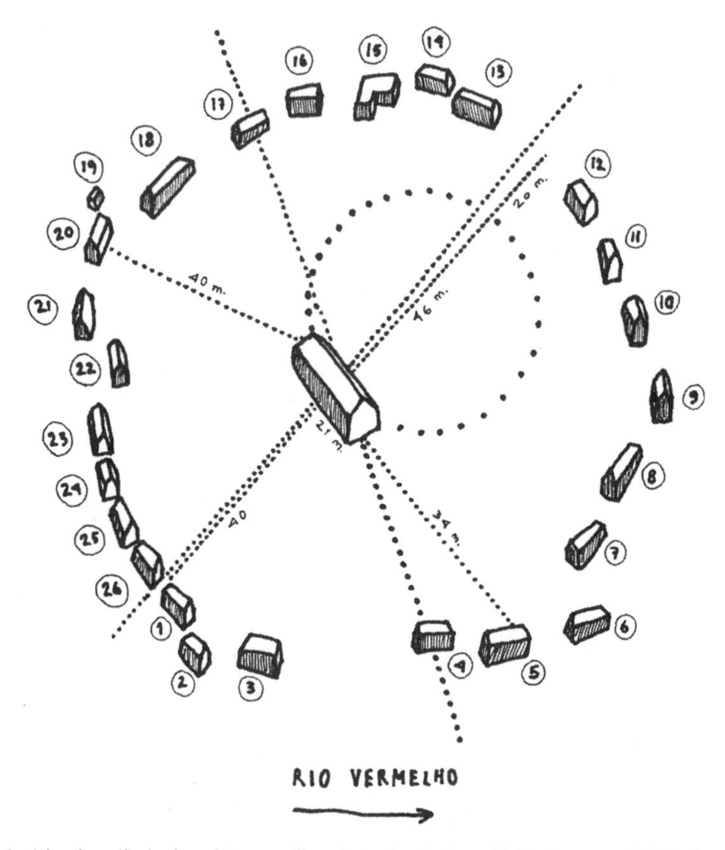

14 Social order reified: plan of Bororo village huts, Brazil, about 1940. Courtesy of Ida Pedersen.

The shape of the village, a living boundary, thus reflected and maintained its social structure; enclosure and dwelling were all of a piece.

The "cartwheel" of huts that formed the boundary of the Bororo village was used not only in permanent settlements, but also in temporary encampments, even those occupied for only one night. This form appeared not only to provide the essential dwelling space of the village, but also to fix relationships and beliefs that would be disrupted, perhaps fatally, in its absence. Changing the physical shape of the boundary was tantamount to destroying villagers' sense of themselves and their place in the cosmos. This discovery was made long before Lévi-Strauss, by the Salesian missionaries who quickly understood that the best way to convert the Bororo was to forcibly rearrange their huts in parallel rows. As Lévi-Strauss wrote, "It was as if their social and religious systems . . . were too complex to exist without the pattern which was embodied in the plan of the village and of which their awareness was constantly being refreshed by their everyday activities."[78]

The boundary of the Bororo village is thus a striking case of an intimate relation between bounding, belief, and dwelling. But one need not look to Neolithic villages to find evidence of this relation. Living in or on the wall, as the story of Rahab illustrates, has marked cities great and small since the beginning of civilization.

In the sixth book of the *Laws*, Plato expounds, through the "Athenian stranger" who does most of the speaking in the work, the roles and functions of city walls. It is safe to say that Plato was generally not fond of walls, viewing them, as did the militaristic Spartans, as a sign of weakness. Plato speaks of walls' tendency to breed "a certain effeminacy in the minds of the inhabitants, inviting men to run thither instead of repelling their enemies, and leading them to imagine that their safety is due not to their keeping guard day and night, but that when they are protected by walls and gates, then they may sleep in safety." However, like Frost's narrator two millennia later, the stranger (and thus, one surmises, Plato himself) appears resigned that walls will nevertheless always be built. Sleeping in safety through the night, he owns, is no small benefit. The question Plato raises is thus a different one: how to build walls well. "If men must have walls," the stranger goes on, "the private houses ought to be so arranged from the first that the whole city may be one wall, having all the houses capable of defence by reason of their uniformity and equality towards the streets. The form of the city being that of a single dwelling will have an agreeable aspect, and being easily guarded will be infinitely better for security."[79] In other words, the city should be its wall, and the wall should become the city. For Plato this is not only a question of military defense: it is equally a matter of aesthetics, or the creation of an "agreeable aspect." For Plato's stranger, the tightly girded compression a wall creates when fused with dwellings is a standard by which to judge the goodness of the city as a whole.

This fusion of wall and city has continued to fascinate urban theorists and architects. In the 1964 Museum of Modern Art exhibition "Architecture without Architects," organized by the Swiss-American architect Bernard Rudofsky, many of the examples were places where the boundary of a settlement was made by the very process of human habitation. One photograph, later included in the exhibition catalog, showed a village in the Moroccan desert whose wall was formed, just as Plato's stranger had instructed, by the combined walls of individual dwellings. "Neither house nor town but a synthesis of both," Rudofsky wrote, "this architecture was conceived by people who build according to their own inner light and untutored imagination."[80]

Rudofsky was most interested in forms and patterns from outside the industrialized world, ones that had been ignored in mainstream architectural education and practice. But he might just as easily have included examples of inhabited walls from the very center of urban civilization. In many cases, the fusion of wall and dwelling occurred just as Plato had described it, with houses and apartments forming all or part of the structural mass of the wall. The arrangement was particularly widespread on the central Italian peninsula, where many towns began as fortresses and the landscape was fragmented into the domains of competing city-states. The town of Capalbio in Tuscany, for example, was enclosed by a wall of apartments that faced the dense interior of the town on one side and a panorama of open fields on the other. The apartments were connected by a public "street" running atop the entire circumference of the parapet. Although these apartments began as residences for sentries, as the fortress grew into a town and the threat from rivals waned, they were gradually turned into permanent residences. The apartments were abandoned in the early twentieth century but have now been reoccupied by merchants and by vacationers attracted to their views and year-round comfort in a climate of cool winters and hot summers.

Like Rahab, who built her house upon the wall of Jericho thousands of years ago, the poor and the marginal of urban society have often tended to find shelter in city walls. Part of the reason lies in urban economics. For most of urban history, land costs in walled cities were highest near the center, where the marketplace, temples, and public buildings were located. This lateral hierarchy of value meant that living on the edge was usually the cheapest way to inhabit the fortified city.[81] Many walled cities throughout history took centuries to fill in their circuits (some never did), leaving a considerable amount of land open for building near the wall. Aside from the temple, the city wall was often the most solid structure in the urban fabric, built to last centuries and withstand sustained assault. Poor residents without access to materials or land therefore used the mass of the wall to reduce the cost of their own residences. Since three-walled structures are cheaper than four-walled ones, it was not uncommon for houses to be built directly against city walls. In other cases people excavated portions of the wall itself to create apartments and

15 Dwelling in defense: street and apartments on city wall, Capalbio, Italy, about 1400. Photograph by Thomas Oles.

dwellings. In the Chinese city of Chengdu, surrounded by a curtain wall fifty feet thick at the base, the poor of the city "not only built dwellings up against the wall, but scooped out caves within it to use for shelter."[82] This practice continued in cities such as Kaifeng into the 1980s, and there were houses backed up against the city wall of Shangqiu in Henan Province until the 2000s.[83] The Great Wall of China was similarly an entire ecology, a kind of linear city housing tens of thousands of soldiers and their retainers.[84] In Renaissance Siena, impromptu structures against and within the city's encircling wall had grown so common by the fourteenth century that statutes were drawn up to tax wall-abutting houses and to legislate where windows might be carved out in order to protect the integrity of the overall structure.[85]

Living in the city wall may have been an act of economic or social desperation, but it frequently put the residents at the very center of history. One natural advantage of dwelling in the wall, as the Siena statutes imply, was to give residents a prospect over both the city and its surroundings, placing them in a unique position to become guards or spies. The account in the book of Joshua suggests that Rahab's lodgings run across the entire mass of the Jericho wall, a position that provides a critical advantage to the Israelites she helps to flee the city. Later she uses this same position to signal the location of her home and save herself and her family from the attacking force.

City walls thus often provided shelter to some of the most vulnerable and marginal people in urban society. But their strategic advantages and prominence in the surrounding landscape also made them the aeries of the powerful. In many of the cities of Mesopotamia, for example, the extensive royal and religious precincts did not lie at the center of the city but rather were built directly against the city wall. At the Assyrian city of Khorsabad, founded in 717 BCE, the *temenos* and royal palace of King Sargon II sat together on a broad plinth formed by the widening of the exterior and interior curtain walls.[86] To those looking up from the outside, the wall of the city and the structure of the palace would have appeared as a single monumental edifice. And in Babylon, the greatest Mesopotamian city, the palace was astride the wall on the northern perimeter, immediately adjacent to the Ishtar Gate, the most important entrance to the city.[87] While placing palaces and temples in this position may appear strange given modern notions of security, it suggests the extent of a city's control over the surrounding territory. The more order reigns in the land outside the city, the more likely the wall is to house priests and kings rather than prostitutes and paupers.

The placement of a palace beside the city gate indicates how important the latter was in any walled city. The wall may have defended the city from attack or conquest at certain points in its life, but its most important daily function was to regulate passage into and out of the city. Gates were therefore usually the most elaborate and opulent parts of any wall circuit. The requirement that gates be manned at all times, both for security and to levy entry tax, yielded another type of dwelling in the wall: the gatehouse or gate tower. Nowhere were these permanently manned structures as elaborate and grand as in ancient Chinese cities. As historian Paul Wheatley wrote, gates were the places "where power generated at the *axis mundi* flowed out from the confines of the ceremonial complex toward the cardinal points of the compass." As such, they possessed a "symbolic significance which . . . was expressed in massive constructions whose size far exceeded that necessary for the performance of their mundane functions of granting access and affording defense."[88] In Rome, each of the sixteen gates in the Aurelian Wall, built in the late third century CE, was flanked by elaborate gatehouses where customs duty was collected. Later, many medieval cities would come to have "societies of the gates," urban guilds whose identity was vested in living close to a particular city gate.[89] As the power of European cities over the surrounding countryside grew during the Renaissance, the natural advantages of living near the gate, equidistant from the central market and neighboring farms, began to attract merchants whose holdings included both urban houses and fields around the city.[90] Such was the case of the Ticino Gate in Milan, whose structure includes private apartments to this day. Similarly, by the fifteenth century in Bristol, rampart towers began to be occupied by merchants.[91] The wall was gradually being transformed from the redoubt of kings to the dwelling of a new urban class.

Gates were more than places where cities asserted the prerogatives associated with their status as loci of trade and commerce. They were places where the mass of the wall became a backdrop for elaborate ceremonies of coming and going, entry and exit, exile and return. Some of these practices persisted long after the physical wall in which the gate was set had vanished. In London, for example, the ritual of presenting the keys to the city preserves the memory of a wall that now exists only in the form of this ancient practice.[92] The third wall of Paris, built in the late nineteenth century long after the need for defensive walls had been obviated, still included a series of elaborate gatehouses at the city's main entrances. And the persistence of gates at the entrance to many modern housing developments, where one must stop and solemnly announce one's business, is a testament to the way passage through a gate can evoke an entire wall as well as the economic or political power that stands behind it. Such practices suggest that walls have always been far more than their stones and mortar. As Frost knew, walls are also rituals repeated day after day, year after year.

Walls Were Ritual

In books 9 and 10 of the *Life of Romulus*, Plutarch recounts the mythical founding of Rome by the orphan brothers Romulus and Remus. He writes that Romulus chose the site of the future city by "a divination from a flight of birds." But disagreement about the number of vultures the two brothers had seen, and thus the propitiousness of the location, led to a quarrel during which Remus mocked Romulus's attempts to mark out the future boundary of the city. "As Romulus was casting up a ditch, where he designed the foundation of the city-wall," Plutarch writes, "[Remus] turned some pieces of the work to ridicule, and obstructed others; at last, as he was in contempt leaping over it, some say Romulus himself struck him, others Celer." After slaying his brother, Romulus "fitted to a plough a brazen ploughshare, and, yoking together a bull and cow, drove himself a deep line or furrow round the bounds; while the business of those that followed after was to see that whatever earth was thrown up should be turned all inwards towards the city; and not to let any clod lie outside. With this line they described the wall, and called it, by a contraction, Pomoerium, that is, *postmurum*, after or beside the wall."[93]

The founding myth of Rome suggests one of the distinctive aspects of Roman culture. Rituals associated with laying out boundaries, many inherited from the Etruscans, played a central role in Roman political and religious life. As the Roman Empire expanded after the first century BCE, the establishment of every new town, garrison, or encampment was what architect and historian Joseph Rykwert called an "anamnesis of imperium," a reenactment of the story Plutarch recounts. The subdivision of conquered land, what the Romans themselves called *limitatio* but today is more commonly called centuriation, was rooted in the same beliefs and involved the same rituals.[94] The work of land surveyors could not proceed without the benediction of an augur, one of a college of official diviners charged with discovering whether the gods approved of a given action. One of these signs was haruspication, or inspection of the entrails of animals slaughtered in the vicinity. To cross what Plutarch would later call the "holy and inviolate place" of the first furrow before these rites had been completed, or to cross in the wrong place as Remus did, was to invite the gods' wrath and undermine the foundation of the settlement. The practices and rituals of Roman subdivision were thus bound up with the deepest beliefs of Roman cosmology, a connection suggested by the name surveyors gave to their field of vision as they looked over an area to be subdivided: *templum*.[95]

Two of the most important events in the Roman festival calendar were thus dedicated to the gods associated with boundaries. The first of these was the Terminalia, the yearly celebration of Terminus, the god who inhered in rural boundary stones. The *Corpus agrimensorum* gives precise instructions on how termini should be set up. It instructs surveyors to follow the practices of "the ancients," who

> would anoint [a boundary stone] and crown it with bands and wreaths. In the ditch where they were going to place it, they sacrificed, and when the victim had been set fire to with a torch, they poured blood into the ditch and threw incense and fruit into it, as well as beans and some wine which it is the custom to offer to Terminus. When the fire had consumed all the sacrifices they placed the stone over the still hot relics and made it sure with the greatest care, reinforcing it roundabout with broken stones that it may stand more securely.[96]

The Terminalia was a reenactment of this ritual on the morning of the twenty-third of February, when neighboring landowners met each other at the boundary stone marking the division between their fields. The most detailed account of the Terminalia comes from the *Fasti*, Ovid's catalog of the Roman festival calendar. "When the night has passed," Ovid instructs his readers,

> see to it that the god who marks the boundaries of the tilled land receives his wonted honor. O Terminus, whether thou art a stone or a stump buried in the field, thou too hast been deified from days of yore. Thou art crowned by two owners on opposite sides; they bring thee two garlands and two cakes. An altar is built. Hither the husbandman's rustic wife brings with her own hands on a potsherd the fire which she has taken from the warm hearth. . . . Terminus himself, at the meeting of the bounds, is sprinkled with the blood of a slaughtered lamb, and grumbles not when a sucking pig is given him. The simple neighbors hold a feast, and sing thy praises, holy Terminus: thou dost set bounds to peoples and cities and vast kingdoms; without thee every field would be a root of wrangling.[97]

The Terminalia was thus a propitiation of a deity, a recollection of the original setting up of the boundary stone, a ritual meeting between two neighbors, and (in its association with imminent spring plowing) a fertility rite. The boundary stone was more than an object in which a mystical presence inhered; it was an altar and the board at which the neighbors broke bread.

The Terminalia had a close corollary in the festival of Lupercalia, celebrated one week earlier on the fifteenth of February. The festival took its name from the Lupercal, the site where the she-wolf was believed to have found Romulus and Remus and the place where its opening rites were performed. The festival saw the initiation of two noble youths into the ranks of the Luperci, a corporation of priests whose shrine was dedicated to the founding of Rome.[98] The neophytes were daubed with the blood of sacrificed goats, during which they were required to laugh. The Luperci then ran a race along the base of the Palatine, following the *pomoerium*, the line of sacred boundary stones that marked Romulus's furrow.[99] As they ran, the priests struck boundary stones and spectators alike with switches.[100] In his description of the Lupercalia, Plutarch tells how young women, in particular, placed themselves in the path of the Luperci in order to be "lustrated."[101] On returning to the Lupercal, the winner of the race was awarded the half-cooked entrails of the sacrificed goats on a willow spit, and a large feast ensued.[102]

By retracing Romulus's original boundary, the Lupercalia both reinscribed the sacred limits of Rome and purified the city and its residents. Furthermore, the active participation of women spectators suggests that the festival, like the Terminalia, was also a springtime fertility rite. These associations, with their provenance in the pagan rituals of the Etruscans, led Christian emperors to condemn the Lupercalia and attempt to ban it. Nevertheless, the festival continued to be celebrated well into the fifth century CE and later, some have suggested, as the modern festival of Saint Valentine.[103]

The systematic, regimented approach to land division and distribution that had characterized the Roman Empire, in which a professional corps of surveyors imposed the same grid pattern on the land everywhere they went, all but collapsed during the Middle Ages. The loss of knowledge and tools related to survey and mapping meant that laying out and recording boundaries became largely a matter of cataloging important local landmarks such as trees, stones, hedges, or ditches.[104] The legal documents listing such landmarks, called "terriers," often ran to hundreds or even thousands of pages. A land grant "map" from early tenth-century England was typical:

On Swinbroc first, thence up from Swinbroc on to rush-slade, from this rush-slade's corner foreagainst Hordwell-way, thence along this way until it comes to the Icknild way, then from these ways upon the old wood-way, then from that wood-way by east Tellesburgh to a corner, then from that corner to a goreacre, thence along its furrow to the head of a headland, and which headland goes into the land, then right on to the stone on ridgeway, then on west to a gore along the furrow to its head, then adown to fernhills slade.[105]

The limits of a parcel were thus a narrative of familiar objects and the stories associated with them. The red oak and spruce in the passage above were not coincidental markers of a boundary with an independent existence that preceded them, but rather the very things that created that boundary as a customary and legal entity. The boundary's validity and force rested in the objects that marked it, and to conceive of it without them would have been impossible.

For this reason, rituals associated with inscribing the location of landmarks in public memory played an extremely important role in the Middle Ages. The Lupercalia and Terminalia continued to have counterparts all across the fringes of the Roman Empire. Many of these rituals were mixed with those of the people the Romans had conquered. In Celtic Britain, for example, villagers struck "besoms," bundles of birch or willow twigs, against boundary markers during the festival of Beltane in early May. Early church historian Henry Spelman suggested that the Roman invaders, finding this ritual similar to their own springtime rites, incorporated Beltane into the celebration of the Terminalia in Britain.[106]

Beginning in the fifth century, such rituals were increasingly appropriated by the Anglo-Saxon church in an attempt to sever their connection to earlier pagan celebrations. By the thirteenth century, the old Roman practices had been thoroughly incorporated into the rite of "processioning," a daylong event held every year during Holy Week or before Ascension, when people would circumambulate the ecclesiastical boundaries of their parish. Practiced both in England and in England's North American colonies, processioning was a major event marked by great fanfare. The leaders of the parish would head up the march bearing crosses and banners, with residents carrying streamers, bells, staves, torches, and candles bringing up the rear.[107] Over the course of an entire day, the column would make its way from landmark to landmark along hedges, ditches, walls, and roads, pausing at "gospel trees," where the parson would read the verse of the day and remind his congregation to respect "bounds and doles."[108] Much like its pagan predecessors, then, processioning was a ritual blessing of land, crops, and people.

Processioning was gradually transformed from a yearly ecclesiastical rite to a secular practice performed once every several years.[109] Known as "beating the bounds" in reference to the thrashing of landmarks with willow wands, the ritual was an important custom in many English parishes, since memory of a landmark's location at a given moment might have legal weight many years later. Because of this, the youngest boys of the community played a key role in the celebration. When the procession arrived at an important landmark, or at a place where the boundary took a sharp turn or was not clearly marked, the boys were struck with wands, thrown over hedges and into ponds, or even held upside down with their foreheads against a stone or tree in order to sear the location of the landmark into their memories.[110] There were rewards too, with cakes scattered at landmarks for boys to collect, further impressing the important day in their recollection.[111] Sometimes the entire procession would perform a collective action as an aide-mémoire. At one early nineteenth-century perambulation in Devon, perhaps in deference to the story of Jericho, the parson declared, "Now we must make a shout here that we may recollect the bounds," at which the group "huzzaed and took off their Hats."[112] The day was capped by an often raucous celebration on the village green. Such revelry was no less important than the solemn parts of the ritual: one sixteenth-century parishioner of New Buckenham in Norfolk noted that he "better remembreth" the location of a boundary "for that he had druncke Beare out of an hande Bell" there.[113]

These rituals show how boundaries were always far more than the objects that marked them. They were sites where people from either side of the boundary came together, either in ancient rituals or at impromptu events, to express the social relationships between them. Beating the bounds, for example, was not just a matter of marking the limits of one's own parish but entailed meeting residents of adjacent parishes performing the very same ritual. And the Terminalia was simply an ancient version of the springtime meeting between Frost's two neighbors.[114] Regardless of the material they were made of, boundaries were stages for collective, shared action, for the unfolding of social life. Finally, they always had less to do with human separation than with human connection.

Walls Were Exchange

A popular novel about the building of a cathedral in medieval England tells how the residents of a growing market town, threatened by a despotic earl and his knights, hastily decide to construct a wall to protect themselves from attack. Under the command of the cathedral's master builder, the townspeople scrape together any materials they have to hand, using timbers and stone designated for the cathedral for a curtain wall and the surrounding earth for an impromptu rampart. Working feverishly, they manage to enclose the entire town in a single night. The result, predictably, is less than ideal: "The stone walls were waist-high, which was not enough. The fences were high but there were still enough gaps for a hundred men to ride through in a few moments. The earth ramparts were not too high for a good horse to surmount."[115]

The earl and his knights are surprised by the sudden appearance of this construction standing between them and the town they were about to attack. The men-at-arms draw up outside the wall, embarrassed and bewildered. They repair to a nearby wood, where they change their plan and regroup. Half the knights then storm the wall while the other half storm the earthen rampart, as the townspeople pelt them with whatever objects they can lay hands on. A few knights manage to climb over the wall and rampart and engage in hand-to-hand combat with the townspeople, who eventually overpower them. The bloodied attackers retreat to their stronghold, while the villagers celebrate their victory even as they know the earl will use this skirmish as the pretext for a new and better planned attack.

The wall described in this novel is a fiction, but it suggests something essential about the social function of walls in general. No matter how high, how hard, how apparently impregnable, walls were rarely built *solely* for the purpose of preventing contact between groups. Instead, they were almost always designed as sites of exchange, even if that exchange took the form of conflict or violence. The wall in this story was hastily built, and its architect, the master builder of the cathedral, knew it would not withstand a sustained assault. What it did do, and did effectively, was to change the rules of the game. The wall set the stage for the conflict between the townspeople and knights in such a way that the former suddenly enjoyed, if not an outright advantage, then a fair chance to defend themselves. The wall did not prevent a conflict from occurring, but it shifted the terms on which that conflict unfolded. In short, the wall staged and regulated social exchange. And in this it resembled real walls throughout history.

Plato objected to city walls on the grounds that they produced an "effeminate" populace, but the builders of Greek cities generally did not share his doubts.[116] The walls of *poleis* were often massive constructions, many having been built on the principles of Cyclopean masonry the Greeks had inherited from the Mycenaeans. Colonial settlements of the Hellenistic period were routinely enclosed by walls to protect them from local populations that were often less than pleased about the new arrivals. The new *polis* of Mantineia, for example, was doubly protected by a wall and a moat created by rerouting the Ophis River.[117] During the Persian wars of the late fifth century BCE, the Delphic oracle predicted that the "wooden walls" of Athens would save the city from invasion, a prophecy Themistocles interpreted as referring to the masts of naval ships. This interpretation proved catastrophically wrong, and the Persians sacked Athens in 480. After the conclusion of the war, the Athenians undertook to build a circuit of new stones around their city, including the famous "Long Walls" that would eventually link Athens to its port, Piraeus.

Despite their essentially defensive nature, however, the walls of Greek *poleis* were anything but impermeable. Instead, they were porous by design, containing "posterns" or small apertures spaced at intervals ranging from thirty to one hundred meters. Posterns differed fundamentally from loopholes or other perforations built into the wall for firing projectiles at an enemy. They were at ground level and were designed to allow troops to sally forth from the city and surprise an attacking force.[118] During the Hellenistic period, when such sallies became common, posterns grew more and more elaborate, their structures integrated into towers from which archers would cover the exiting infantry below.[119]

Posterns thus offered a significant strategic advantage to the force defending a city from attack. But if the possibility of emerging from the wall at any moment sometimes improved the chances of those inside, posterns also offered attackers numerous opportunities to enter the city and overwhelm their opponents. This too was strategic, since posterns often opened into courtyards inside the wall where an attacking force might find itself trapped. But insofar as it involved admitting the opposing force, the strategy was risky. Even when defenders sallied out of the enclosure, the opposing force had a brief but crucial advantage, since the narrow openings forced defenders to exit single file.

To modern eyes the postern appears to be, if anything, a strategic oversight rather than a strategic advantage. Why go to the trouble of building a masonry wall and then undermine it with holes any determined enemy can pass through? But posterns make no sense only if one assumes that the function of a wall is to completely separate one group from another. Greek city walls were not designed to isolate the cities they surrounded, but rather to set the terms of exchange with the people

16 Channel of communication: postern gate in citadel, Mycenae, Greece, about 1350 BCE. Courtesy of Erich Lessing/Art Resource, New York.

and groups outside them. Their builders understood that no barrier was impregnable and that any enemy would eventually find its weakness, even to the point of tunneling underneath. They therefore preemptively designed this weakness in the structure of the wall itself. Enemies knew the risks of penetrating the wall, and defenders knew the risks they took in providing the chance to do so. Greek walls were thus distinguished by carefully regulated permeability in both directions, permeability that provided "crucial support for ongoing strategic interaction."[120] Posterns also served eminently practical purposes during times of peace. Since they gave easy access to the countryside, residents and visitors often used them for everyday transactions between the city and its surroundings, in lieu of the more formal city gates.[121]

The postern is but one characteristic of settlement walls in general. However impregnable and impressive they may have appeared, walls were almost never built to seal people off from their surroundings. Instead, far more often, walls embodied the relationship between the two. And the stories above notwithstanding, history demonstrates that this relationship is not always, or even most often, one of violence.

In the first volume of the *Histories*, Herodotus describes the construction of the massive outer wall of Babylon, a city with "magnificence greater than all other cities of which we have knowledge," built in the sixth century BCE during the reign of Nebuchadnezzar II:

> There runs round it a trench deep and broad and full of water; then a wall fifty royal cubits in thickness and two hundred cubits in height. . . . As they dug the trench they made the earth which was carried out of the excavation into bricks, and having moulded enough bricks they baked them in kilns; and then afterwards, using hot asphalt for mortar and inserting reed mats at every thirty courses of brickwork, they built up first the edges of the trench and then the wall itself in the same manner: and at the top of the wall along the edges they built chambers of one story facing one another; and between the rows of chambers they left space to drive a four-horse chariot.[122]

Herodotus calls special attention to the "hundred gates, all of brass, with brazen lintels and side posts" set into this great structure. The most magnificent of these was the city's northern entrance, dedicated to the goddess Ishtar, located at the terminus of the Processional Way that ran through the city. The Ishtar Gate was approximately forty feet high and embellished with the glazed reliefs of nearly six hundred dragons and bulls, arrayed in thirteen ranked tiers on its facade. Robert Koldewey, the German archaeologist who excavated Babylon in the early twentieth century and brought the remains of the gate to Berlin, called it "the most striking ruin of Babylon and . . . of all Mesopotamia."[123] It was every bit as great a wonder as the royal palace and hanging gardens just inside it.

But the real activity did not happen inside the Ishtar Gate or any other gate. Babylon lay on the major trade route between the Persian Gulf and the Mediterranean, and just outside the city's entrances there grew up secondary settlements where traders from distant lands could gather and exchange wares without having to pay entry tax. Koldewey showed five such suburbs, what the Babylonians called a *karum*, near the major gates in the outer wall. These settlements were the site of bustling marketplaces and temples, some of them nearly as important as those inside the wall.[124] Indeed, many of these early suburbs became so vital to the life of the city that they were later enclosed by their own walls.[125]

The layout Koldewey described was not unique to Babylon but has marked nearly every walled city throughout history. When in the late third century CE the Roman emperor Aurelian ordered the construction of a new wall that would take in as much of the city as possible, the *suburbia* that grew outside its sixteen gates emerged as critical points of exchange between the capital and the far-flung regions of the Empire.

17 Axis of exchange: Ishtar Gate and Processional Way, Babylon, Mesopotamia, about 600 BCE. Courtesy of Ida Pedersen.

The *suburbium* housed many buildings and activities that were essential to the daily functioning of the city but considered noxious within its bounds, such as magazines and depots, livestock pens, and tanning operations.[126] Roman legions were not allowed to enter the city, so the *suburbium* was also the place where their various needs were satisfied; suburbs were generally associated with prostitution. Like the Babylonian *karum*, then, *suburbia* took on their own distinct life, one oriented both outward and inward, toward physically essential but socially marginal activities. This sense of difference persists to this day: modern residents of Rome continue to distinguish the city proper from areas *fuori porta*, "outside the gates."

The Aurelian wall, like many urban walls throughout history, was an emblem of waning imperial power, constructed in response to the growing threat of barbarian invasion from the north. Beginning in the fifth century CE, the vacuum created by the breakdown of Roman political authority in Europe began to be filled instead by local potentates, ruling from a network of permanent garrisons scattered across the continent.[127] These strongholds went by names derived from either the Latin *castrum* (Italian *castello*, English *castle*) or the Low German *burgus* (French *bourg*, Italian *borgo*, English *borough*). Historian Henri Pirenne described *burgi* as "walled enclosures of somewhat restricted perimeter, customarily circular in form and surrounded by a moat."[128] In the center of a typical *burgus* stood a strong tower and keep, the last redoubt for the prince and his retinue in case of attack. But the *burgus* was also a refuge for the entire local population. "Though primarily built for the protection of the tenant (lord) and his dependents, their flocks and their herds," historian Alfred Harvey wrote, the *burgus* "also performed its part in the general defense of the neighborhood."[129] Beginning in the ninth century CE, "countless burgs were built by dukes, counts, and margraves . . . for protection against Norsemen, Magyars, Slavs, Saracens," as well as against rival princes.[130] English historian Frank Stenton estimated that the county of Wessex alone contained thirty-one "boroughs," meaning no rural village was more than twenty miles from safety in times of war.[131] At the same time, little more than a century later, invading Normans would use *burgi* as strongholds from which to administer and control the newly conquered land of Britain.

But the *burgus* not only was a locus of defense or power over an area; it was also a place of permanent habitation and use, of bustling activity even in times of peace. In particular, like the first cities and villages, it was effectively a granary, promising sustenance or at least survival during poor harvests. The *burgus* was also a place of monetary profit. As trade revived across Europe after the tenth century CE, merchants began to stream toward the literally captive markets of *burgi* as places of commerce. But there was not enough space in most fortresses to accommodate these new arrivals. Itinerant merchants therefore settled just outside the *burgus*, in the shadow of its wall, where they built a *foris burgus* or "outside burg," a word that would later yield the French *faubourg*. These extramural suburbs were also called *novus burgus* to distinguish them from the *vetus burgus* inside the wall. In the Netherlands and England, the Roman word *portus*, "gate," was gradually adopted to denote these areas, the origin of the modern "port."[132]

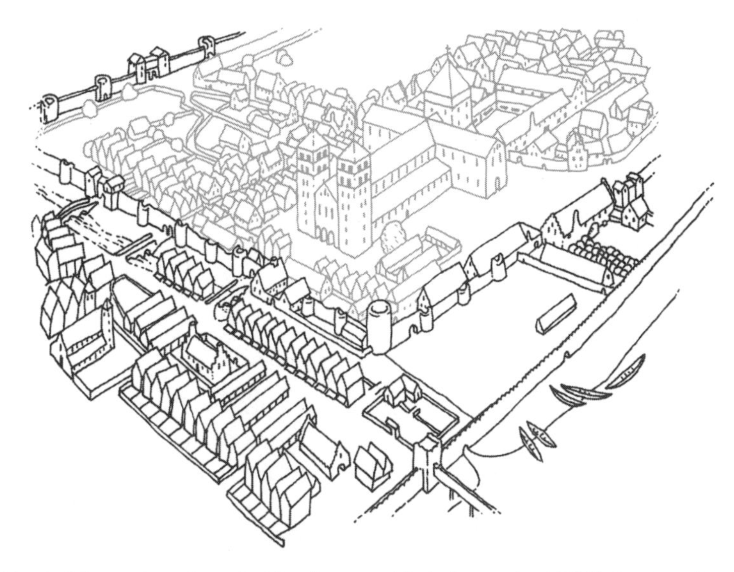

18 Fount of the modern city: walled Strasbourg and its *faubourg,* about 1200. Courtesy of Ida Pedersen.

The inhabitants of the *foris burgus* were people whose livelihood depended on exchange. "The military presence," historian A. E. J. Morris wrote, "generated immediate service industry activity and a produce market would have soon become established to provide for daily needs of both the military elite and the serf community."[133] The people who provided these services gradually came to be called *burgi,* in contrast to the *castellani* living inside the wall. Huddled just outside the gate, protected by the looming presence of the fortress but still exposed to frequent banditry, these "burghers" rapidly emerged as an economic engine. What had begun as ad hoc settlements gradually acquired independent legal status, as an economy of exchange began to replace the subsistence economy that had dominated the early Middle Ages. Thus, as Lewis Mumford wrote, "a new class got protection against theft and arbitrary tribute, and began to settle down permanently, just outside the walls."[134] As this class grew in economic importance, many *faubourgs* were themselves later surrounded by masonry walls or wooden stockades, creating secondary enclosures similar to those Koldewey discovered at Babylon.[135]

Like the Roman and Babylonian suburbs before it, the medieval suburb gradually took on its own distinct culture at the margins of urban life, based on exchange between *castellani*, residents of the neighboring countryside, and travelers from distant lands. In *The Canterbury Tales*, Geoffrey Chaucer captured the sometimes seedy aspect of life outside the wall:

Wher dwellen ye, if it to tellen be?
In the subarbes of a toun, quod he,
Lurking in hernes and in lanes blinde,
Wheras thise robbours and thise theves by kinde
Holden hir privee fereful residence.[136]

The extramural zone also came to house lepers and victims of plague. Nineteenth-century Italian writer Alessandro Manzoni described the Milan lazar house of the seventeenth century as a "rectangular enclosure, almost square in fact, outside the city proper, to the left of the East Gate, and separated from the city wall only by the width of the moat."[137] But the marginal location of the suburb and its separation from the main part of the city also allowed what today is considered urban culture to develop and flourish. In Elizabethan London, for example, Chaucer's "lanes blynde" housed not only robbers and prostitutes, but also Shakespeare's Globe Theatre.

The *faubourg*, the *suburbium*, and the *karum* were essential threads in the city's social and economic fabric, their particular culture due to the nature of interactions banished from the city but sanctioned just outside its gates. These interactions created the dynamic social and economic relationships that would increasingly come to characterize cities in the modern era.[138] In many ways the roots of the modern city lie not in the central square, but in the shadow of the wall.

—

The image of a hilltop fortress, standing aloof above its fields and woods, expresses perhaps better than any other the age-old connection between walls and domination. Since the beginning of civilization, walls were used to assert power, conquer people and territory, and solidify social arrangements rooted in unequal access to resources, whether of grain, tools, or knowledge. But the power of the few over the many is only one part of the story of what walls were. The endurance throughout history and prehistory of fences, hedges, ditches, and earthen banks, objects made and sustained by ongoing human labor, can never be fully explained by an account focused exclusively on power and violence. To give such an account is to remember only the part of the story in which the walls come tumbling down and to ignore the Rahabs who lived on them. It is to forget that, from Jericho to Skara Brae to the *burgus*, walls had as much to do with sustenance and increase of human collectives as with their control or destruction. They were never either/or but always both/and—both coercion *and* dwelling, force *and* belief, violence *and* nurture.

If there is one thing that unifies the places and times discussed in this chapter, it is that wherever and whenever they were built, walls were by necessity rooted in the human economy and natural environments where they were made. A ditch dredged from the alluvial soil of Babylon yielded the mud bricks that made its enceinte; the wall of the medieval fortress was built from the tufa carried from the nearest quarry; and a hedge in Cornwall literally grew out of the earthen bank beneath it. The wall was the natural landscape transformed by human ingenuity and labor. It was unavoidably and palpably of its place.

Viewed this way, it becomes absurd to think of walls as objects that "divided" people or groups. Like everything people have made, walls were always bound up in social relationships. This might mean the trade between ancient cities, the battle between rival princes for control over the land between their fortresses, or the annual meeting of Roman farmers at a boundary stone. But the idea remains the same: walls did not enforce *divisions* but rather staged *interactions*, of which violent conflict was only one kind. Throughout history and prehistory, walls were expressions of the people, beliefs, and rituals that obtained in the places where they were built. They were always primarily socially constitutive rather than socially corrosive. Walls did not divide landscapes; they created them.

This socially constitutive aspect of walls was to be challenged after the seventeenth century, when new notions of the political sovereignty of the nation and the economic sovereignty of private property spurred the development of a radically new technics of enclosure. These developments would fundamentally change the ways boundaries were made and understood. From Old World to New World, from country to city, the advent of the modern subdivided landscape was at hand—and with it the impoverishment of the wall.

THREE

Constructions of Sovereignty

In the summer of 1830 the villages surrounding Otmoor, near Oxford in England, were transformed into a battleground. Residents of the villages had long used the moor for grazing, fuel, and food, uses protected by an elaborate system of common rights reaching back many centuries. In the late eighteenth century, however, the well-known agricultural reformer Arthur Young had declared the soil of Otmoor ideal for farming, and landowner Alexander Croke had proposed that the moor be drained and "enclosed," or divided into private holdings.[1] Public notices announcing the plan were torn down by angry villagers, and the local earl blocked Croke's bill in Parliament. "As Lord of the Manor," he explained, "I should feel it my duty to protect the cause of the *poorer* inhabitants in preference to those of the *richer* who need no such protection."[2]

For the next three decades, the enclosure of Otmoor was delayed by disagreements among landowners in the district and remained primarily a local matter. In 1830, however, the conflict exploded into what would become known as "the Revolution of Otmoor." When a cut was made in the river Ray to ready the moor for enclosure, the upstream banks broke and flooded prime agricultural land. Tenant farmers whose crops were inundated attacked and destroyed the embankment of the new cut. Twenty-two men were indicted, but their protest had emboldened others opposed to the enclosure. Groups from the villages around Otmoor began routinely breaking fences erected to mark new holdings. The largest single protest occurred in early September, when a crowd of about five hundred villagers wrecked fences across the common, to the cry "Otmoor for ever!" Forty-two arrests were made, but the prisoners were freed by a popular riot.

As the months passed, sporadic attacks grew increasingly methodical. Throughout 1831 and 1832, each night "crowds of between one and five hundred ventured on to the moor with blackened faces, frequently disguised as women, and carrying pitchforks, bill-hooks, and even guns."[3] Their goal was to destroy new fences, hedges, and bridges built the previous day. Opponents to the enclosure were supported by many parish constables. One overseer in the village of Charlton provided beer at the local pub and instructed residents to "go into Otmoor, and cut all down if you can"; a box labeled "Otmoor Subscriptions" collected funds for these protests. One landowner who stood to benefit from enclosure complained that "the lower classes of Farmers and the poor in the whole neighbouring Country take part" in the fence raids.[4] However, by the mid-1830s this alliance between farmers and tenants had broken down, the former "rediscover[ing] a deep distrust for the excesses of popular activity," the latter "profoundly suspicious of the farmers' overweening self-interest."[5] At the same time, landlords were increasingly adopting techniques similar to those of the protestors, building new fences under cover of night and with the protection of a police force convened expressly for the purpose. Opposition gradually withered, and the common was finally enclosed in 1835.

Lasting more than forty years, the conflict over Otmoor was the longest in the history of parliamentary enclosure. Seen from the vantage point of the poor and landless, the enclosure was a watershed event. In the space of a few decades, what had been a common resource for the surrounding villages—providing food, fuel, and building materials to those without land—was transformed into a landscape of private plots controlled by the largest landowners in the district.[6] The new fences and hedges on Otmoor thus embodied the ascendancy of one class of rights, those of *property*, over another, those of *use*.

The enclosure of Otmoor is a striking illustration of the way ancient activities of planting hedges and building walls were drafted, at the beginning of the nineteenth century, into the service of a political and economic program aimed at transferring common resources into private ownership. Yet the protests, however dramatic, were nearer the culmination of that program than its beginning. The stage for the transformation of the wall had been set long before the first fence cutters set foot on the moor.

Marking Property, Making the State

In 1690 John Locke published the *Second Treatise of Government*, among the most influential works of political philosophy in the English language. The *Treatise* is a cornerstone of modern political and economic liberalism and has been used in the three centuries since its appearance both to support governments and to justify revolutions against them. More than any other text, it is the reason Locke can be considered "the most influential philosopher of modern times."[7]

The *Treatise* is concerned with two main questions: the foundation of the state, and the nature and origins of property. Locke viewed these questions as two parts of a single philosophical inquiry into the extent and end of civil government.[8] Before government, he argued, people had lived in a condition where they enjoyed "perfect freedom to order their actions, and dispose of their possessions and persons, as they think fit, within the bounds of the law of nature, without asking leave, or depending on the will of any other man."[9] Locke called this condition the "state of nature," a phrase he borrowed from Thomas Hobbes, whose *Leviathan* had been published forty years earlier. Hobbes took a dark view of human life in the state of nature, famously describing it as "solitary, poor, nasty, brutish, and short."[10] Lacking a sovereign, Hobbes believed, people were bound to descend into violence as they pursued their own interests without regard for those of their fellows. Government was therefore a "social contract" in which people agreed to relinquish their freedom and submit to this sovereign—the Leviathan—in order to protect themselves from one another and thereby to ensure the possibility of general increase.

Locke's vision was less pessimistic than Hobbes's. The state of nature was not a nightmarish hypothetical condition, but rather the actual situation that had prevailed among people before the invention of "politic societies." Locke held that people were "naturally inclined to seek communion and fellowship with others" and, like Hobbes, viewed government as a contract concluded "to supply those defects and imperfections which are in us, as living single and solely by ourselves."[11] But Locke imagined the source and legitimacy of that contract very differently than Hobbes did. Whereas for Hobbes the social contract was rooted in people's need to preserve their own bodies, for Locke its original and chief function was to protect *property*.

One of the main premises of the *Second Treatise* is that "men, once being born, have a right to their preservation, and consequently to meat and drink, and such other things as nature affords for their subsistence," and that "the earth, and all that is therein, is given to men for the support and comfort of their being."[12] At the same time, Locke argued, there must necessarily exist a means to "appropriate" this munificence "before [it] can be of any use, or at all beneficial to any particular man."[13] Locke thus viewed the right to property as deriving directly from the original condition of people in the state of nature. The question was how this right first came to be exercised, or how the chain of property had started.

The first and most basic form of property, Locke held, was the individual's ownership of his own body. In the state of nature, every person was "absolute lord of his own person and possessions, equal to the greatest, and subject to no body."[14] The view that all people were originally sovereigns was a radical notion at a time when many, if not most, people across the European continent still belonged, by law or by custom, not to themselves but to lords, earls, counts, or kings. This original sovereignty was the source from which all other forms of property and possession derived. Property, Locke argued, was the result of commingling one's labor, owned absolutely by dint of the individual's sovereignty over his body, with the raw material of nature. "Though the earth, and all inferior creatures, be common to all men," he wrote,

> yet every man has a property in his own person: this no body has any right to but himself. The labour of his body, and the work of his hands, we may say, are properly his. Whatsoever then he removes out of the state that nature hath provided, and left it in, he hath mixed his labour with, and joined to it something that is his own, and thereby makes it his property. It being by him removed from the common state nature hath placed it in, it hath by this labour something annexed to it, that excludes the common right of other men.[15]

In the sense in which Locke uses it here, nature is a "common," a boundless munificence granted by God. Property comes into being when people, through their own labor and use, alienate some part of this bounty.[16] For Locke the process was essentially the equivalent of fencing or walling a piece of common land and bringing it into cultivation, making it productive. "As much land as a man tills, plants, improves, cultivates, and can use the product of," he wrote, "so much is his property. He by his labour does, as it were, inclose it from the common."[17] Once land has been alienated in this fashion, it ceases to be part of a general inheritance, and the rights of the encloser necessarily exclude the "common right" of others. However, "if either the grass of his enclosure rotted on the ground, or the fruit of his planting perished without gathering, and laying up, this part of the earth, notwithstanding his enclosure, was still to be looked on as waste, and might be the possession of any other."[18] Keeping a piece of land enclosed was thus essential to retaining it as one's own possession. Constructing a fence or wall around a plot of land did more than merely symbolize property: it *created* it.

For Locke, the natural right to property preceded all other forms of sovereignty. However, this had created a fundamental tension, since individual claims to the common inevitably came into conflict. The law of nature, according to Locke,

> serves not, as it ought, to determine the rights, and fence the properties of those that live under it, especially where every one is judge, interpreter, and executioner of it too. . . . To avoid these inconveniences, which disorder men's properties in the state of nature, men unite into societies, that they may have the united strength of the whole society to secure and defend their properties, and may have standing rules to bound it, by which every one may know what is his. To this end it is that men give up all their natural power to the society which they enter into, and the community put the legislative power into such hands as they think fit, with this trust, that they shall be governed by declared laws, or else their peace, quiet, and property will still be at the same uncertainty, as it was in the state of nature.[19]

In other words, the social contract is concluded in order to preserve the right of each person to "inclose" some part of the common. In forming governments, people agree to cede "absolute lordship" to a sovereign who will protect their property and arbitrate competing claims among them.[20] At the same time, Locke argued, people retain the right to revoke their consent to be governed should the sovereign fail to perform these functions, a position that was to provide, less than a century later, the justification used by the North American colonies for their own independence.

In theory, the rights Locke described extended to all human beings. Every person, no matter how lowly, was originally "equal to the greatest." But one only needed to look across the landscape of England in the late seventeenth century to understand that this did not remotely describe the social conditions of the country. In practice, rights were anything but equal. A small number of men owned land while large parts of the rural population were bound to estates through ancient systems of tenancy and obligation.[21] Locke believed that only landowners should hold the franchise; men of property were the only admissible signatories to the social contract, and they alone were entitled to change its terms.

The social contract was thus not a general agreement concluded by all people, but rather a pact between two kinds of sovereigns: on one side property owners, on the other the monarch.

Locke wrote the *Treatise* at a time of rapid transformation. England had only recently emerged from the upheavals of the Civil War, Commonwealth, and Restoration. In 1688 Parliament had installed William of Orange on the throne, in the so-called Glorious Revolution. The power of the king was now held in check by the power of Parliament, a representative body composed of, and elected by, male landowners. In this context Locke, a member of the rising class of Whig aristocrats, was not simply describing an ideal relationship between ruler and ruled; he was constructing a philosophical justification of that relationship as it had actually developed in his own society. In the words of political scientist Richard Marens, Locke created a "fiction of government originating by voluntary agreement for the express purpose of protecting pre-governmental property rights, a position that would deny legitimacy to any government that threatened those very rights it was allegedly created to protect."[22]

Throughout the *Second Treatise,* Locke employed the metaphor of the fence to describe the agreement between property holders and the monarch. Not only was the act of enclosure capable of bringing about property ex nihilo; the fence also stood for the fundamental relationship between ruler and ruled in the social contract. Kings were "guards and fences to the properties of all the members of the society, to limit the power, and moderate the dominion, of every part and member of society."[23] In their turn, the property owners at whose pleasure kings ultimately served were "fences" against royal transgression or excess. But the fences that run through the pages of the *Second Treatise* were not only a symbol. They were real things of wood and stone, and even as Locke wrote, they were covering the landscape of middle England.

Owners' Little Bounds

Forty years before the appearance of the *Second Treatise,* agricultural economist Cressey Dymock published a pamphlet titled "A Discovery for New Divisions, or Setting Out of Lands, as to the Best Forme." The pamphlet was published in the form of a personal letter addressed to Samuel Hartlib, Dymock's associate and friend. In the letter, Dymock lamented that "all or most part of the Lands, Lordships, Manors, Parishes, Farmes, and particular Grounds, or Closes in England are not . . . set out in any good Forme; too much of England being left as waste ground in Commons, Mores, Heaths, Fens, Marishes, and the like, which are all Waste Ground." He acknowledged that some of these wastes were "being made a little better use of than others," but insisted that all were "capable of very great Improvement, as not now yielding . . . the one fourth part of that profit either to private or publique, which they are respectively capable of."[24]

By way of illustration, Dymock proposed two hypothetical plans for draining and reclaiming the marshlands of East Anglia. The first was a circular plan of "one Entire Lordship, or Manor-house, with its proper Demains" of "100, 200, or 300 Acres." The plan showed the mansion at the exact geometrical center of the circle, which was divided into four equal quadrants by drainage ditches. Immediately adjacent to the house lay kitchen gardens, orchards, and dairies, beyond which arable fields extended outward toward the encircling main channel. At the far reaches of the plan lay pastures for "Lean Sheep, dry Cows, or young Beasts."[25] The second, far larger plan showed a grid of "16 great Farms, conteining 100 Acres apiece, and 16 lesser Farms, consisting of 25 Acres apiece," whose limits were marked by roads, ditches, and two perpendicular "great Drains."[26] At the bottom of this network of boundaries lay a "main River" into which the whole area would drain.[27] As in the first proposal, the manor house stood at the exact center of the grid, whose divisions replicated in miniature the larger plan, with a smaller but still substantial farmhouse in each.

Dymock argued that such schemes, by bringing marginal lands into production, would inevitably advance "the Common good."[28] Yet by the seventeenth century, there were few if any places where they might be implemented without fundamentally altering the social as well as the physical landscape. Many of the "mores, heaths, fens, marishes, and the like" that Dymock decried were not "waste ground" at all, but rather commons actively used and managed by rural communities. The scale of land reorganization and reconstruction he proposed would have been feasible only for the wealthiest landowners, men with the political and economic power to transfer common land into private ownership. The geometrical boundaries and central manor houses that Dymock envisioned would have been an unmistakable expression of this sovereignty.

Dymock's plans thus not only were an argument for improving agriculture; they were diagrams of a social hierarchy, bearing an uncanny resemblance to those being produced at the same moment in Europe to depict the power of princes and kings. Lest the implication escape his readers, Dymock spoke directly to the future occupant of the great

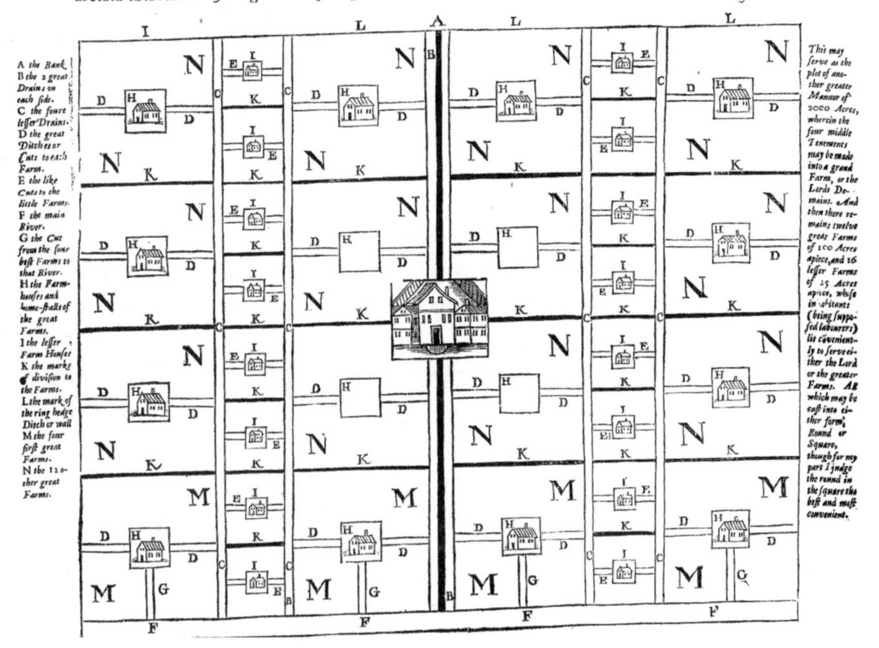

19 Marking boundaries, making property: Cressey Dymock's ideal reclamation, East Anglia, England, 1653.

house: "Finally, here your house stands in the middle of all your little world . . . enclosed with the Gardens and the Orchards, refreshed with the beauty and odour of the blossoms, fruits and flowers, and the sweet melody of the chirping birds, that again encompast with little Closes, that all young, weak, or sick Cattle may be fostered under your own eye without losse or inconvenience, and all bound together as with a girdle."[29]

Dymock presented his proposal as a project of national significance, one that would inevitably redound to the benefit of all England. Reclaiming and subdividing land, even when it served private interests, would result in clear *public* good. The geometrical ditches and canals that would both drain the land and mark the domains of landowners would embody this dual benefit. In the seventeenth century, it would have taken considerable technical and economic resources to implement such ambitious schemes. As historian John Dixon Hunt has noted, "It was not always possible, owing to topographical or financial exigencies, to make this hierarchy of cultural control over territory visible on the ground."[30] Dymock's plans for East Anglia thus remained largely in the realm of speculation. Within a century, however, diagrams very similar to those he proposed would begin to assume real form in the landscape. Ditches, hedges, fences, and walls were about to become not mere symbols of property but, as Locke would imagine several decades later, the very instruments of its creation.

Cressey Dymock was only one of many "improvers" roaming the fields of England from the seventeenth century to the nineteenth, gentlemen of education and property committed to modernizing agriculture through reclamation, introduction of new crops and varieties, and conversion of land from grain production to relatively more profitable animal husbandry. Men such as Dymock, William Marshall, and most indefatigably, the polymath Arthur Young argued that these changes were both an economic and a moral imperative for the nation. In making this argument they used language very similar to that of Locke in the *Second Treatise*. In his 1799 *General View of the Agriculture of the County of Lincoln*, for example, Young surveyed with satisfaction "a large range which formerly was covered with heath, gorse, etc., yielding in fact little or no produce, converted by enclosure to profitable arable farms . . . and a very extensive country all studded with new farmhouses, offices, and every appearance of thriving industry."[31] Improvement, in other words, was largely synonymous with enclosure and private ownership.

Young and his allies identified two main impediments to the changes they advocated. The first was open-field cultivation, the subsistence form of agriculture brought by the Saxons that had characterized many English villages throughout the Middle Ages. In open-field systems, three or more great fields were plowed cooperatively in strips belonging to individual households, and fences were moved over the course of the year to allow livestock to graze on fallow land and on stubble after the harvest. Improvers never tired of inveighing against open fields as inefficient and wasteful, since strips belonging to a single household were often spread over a large area. Young called those who tilled them "Goths and Vandals."[32]

The second object of the improvers' criticism was commons such as Otmoor. Dymock and his successors saw in these "waste grounds," primarily marshes, woodlands, or upland moors, indefensible anachronisms hindering the economic development of the nation. And indeed many improvers compared their campaign against commons to a war against a foreign power. "We have begun another campaign against the foreign enemies of the country," wrote Sir John Sinclair in 1803. "Let us not be satisfied with the liberation of Egypt, or the subjugation of Malta, but let us subdue Finchley Common; let us conquer Hounslow Heath; let us compel Epping Forest to submit to the yoke of improvement."[33] This great national project was a matter of subduing not only the commons themselves, but also the "commoners" who used and depended on them, people the improvers viewed as nothing short of lawless savages. Young, for his part, professed to know "nothing better calculated to fill a country with barbarians ready for any mischief than extensive commons."[34] Commoners in this view were an insidious internal enemy that threatened to eat away the very fabric of the nation.

To vanquish this enemy, improvers advocated a program of radical land reform overseen and enforced by Parliament. Between the mid-eighteenth and mid-nineteenth centuries, this program, known collectively as parliamentary enclosure, affected approximately 7 million acres of land, or nearly one-quarter the area of England.[35] Parliamentary enclosure took place in two distinct waves corresponding to the two conditions improvers saw as impediments to modernization. The first wave, which lasted approximately from 1750 to 1780 and took place during a depression in grain prices, affected approximately 4.4 million acres of open-field arable land, or "champion," in England's Midland counties, most of which was converted to closed pasture. The second wave, between 1793 and 1815 amid the price inflation of the Napoleonic Wars, reclaimed, divided, and transferred into private ownership approximately 2.3 million acres of common "waste."[36] As the history of Otmoor shows, it was this second phase that provoked the most bitter resistance in many rural communities.

There was nothing new about enclosure per se. The landscape of England had been divided by banks, hedges, walls, and fences for millennia, the open fields having only partly replaced older farming systems, like those of Cornwall, in which pasture and arable land were permanently enclosed. What is more, England had been undergoing a slow but steady re-enclosure since the Stuarts in the early sixteenth century; one historian has estimated that three-quarters of the arable land in the country had *already* been enclosed by 1760.[37] Writing of this earlier wave of enclosure, during which many landowners evicted their tenants to create more space for raising sheep, Thomas More lamented that "each greedy individual preys on his native land like a malignant growth, absorbing field after field and enclosing thousands of acres in a single fence."[38] Parliamentary enclosure was thus not the first, but closer to the last in "a long series of processes in which rights to the use of land . . . were defined increasingly clearly and carefully."[39]

This, however, should not obscure the novelty of parliamentary enclosure with respect to the methods and technics of land division that prevailed before it. Throughout much of the Middle Ages, creating fields with hedges or walls was a small-scale undertaking that generally took place as a result of ad hoc agreements among villagers and landowners. So-called piecemeal enclosure "involved a series of private agreements which led to the amalgamation, through purchase and exchange, of groups of contiguous open-field strips, and their subsequent fencing or hedging."[40] By the end of the eighteenth century, such local and piecemeal decisions were systematically and decisively transferred to a centralized body, the Westminster Parliament. The result was to reshape both English society and the English landscape.

An enclosure by parliamentary act began with an exchange of letters or an informal meeting among the landowners in a given parish. The largest owners were usually the strongest advocates, but it was not uncommon for smaller owners to endorse an enclosure to consolidate or expand their holdings.[41] Four-fifths approval among landowners was generally required to enclose a parish; in practice this could be as few as three or four people.[42] A petition was then drafted requesting the right to introduce a bill of enclosure in Parliament. Upon passage of the bill, an enclosure commissioner was appointed by Parliament, or by the landowners themselves. The main qualification for assuming this role was status as a man of "good standing" who could be trusted to distribute poor and rich land equitably among owners.[43] Commissioners were not allowed to reside in the parishes they were charged with enclosing, a policy nominally designed to ensure fairness but that also prevented sympathy toward local opponents.[44]

Once appointed, the commissioner engaged a surveyor to record the topography and existing field boundaries of the parish. This survey replaced all earlier legal records, most of which were based on human memory or written terriers. Then, in a meeting that lasted several days and often took place at a remote location, particularly when local opposition was strong, the parish was reallotted and the new parcels drawn up on an "enclosure award" map. This way of laying out land was fundamentally different from those used in the Middle Ages and had not been employed since the Romans left Britain in the third century

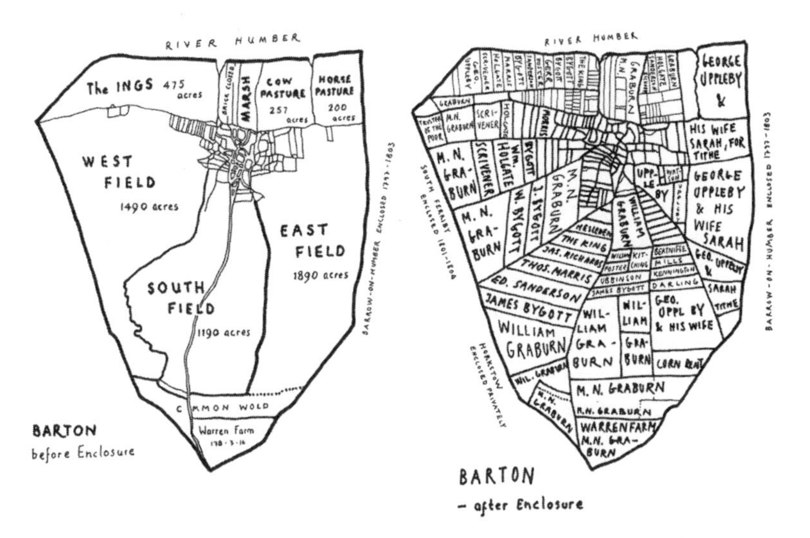

20 New world of walls: parliamentary enclosure award map for Barton, England, about 1800. Courtesy of Ida Pedersen.

CE. Whereas a medieval terrier had been a *record* of fields defined by meaningful landmarks, an enclosure award map was a *projection* of new boundaries where none yet existed. Only landowners were present at the meeting where the award map was drafted, and, unlike the piecemeal enclosures and "enclosures by agreement" of an earlier era, which had been arbitrated and sometimes reversed by manorial courts, there was no contesting or appealing the commissioner's boundaries. They had the force of national law.[45]

These boundaries often fundamentally reconfigured the landscape. Commissioners either consolidated disparate holdings into larger and more regular fields, many of them exceeding fifty acres, or subdivided previously open areas such as heath and moor. In both cases they tended to lay out squarish parcels with long straight edges that ignored natural topography, their construction aided by rapid improvements in survey instruments and standard measuring chains.[46] From the air the ideal parliamentary enclosure would have resembled nothing so much as the rational grid of farms Dymock envisioned more than a century earlier.

But it was not only the shape or size of fields that distinguished parliamentary enclosure from its predecessors. It was the way the *boundaries* of these fields were made, and the stuff of their composition. A parliamentary act invariably included detailed instructions about how and when new holdings should be physically enclosed. Landowners were generally given three to nine months to mark their property with fences, hedges, or walls, and those who failed to do so were subject to heavy fines. These requirements applied to all landowners whether they had opposed or supported the enclosure, and the costs of fencing were to be borne by the landowner alone. In the first phase of enclosure, of open-field arable land, hedges were more common; in the second phase, of commons and "waste," which tended to be poorer in timber resources, stone walls were more often used. The result was an immediate and dramatic growth in demand for fencing materials such as posts and rails, particularly acute in areas devoid of woodland.[47]

The fencing requirements also led to considerable simplification in the structure and diversity of hedges, which remained the most common means of fencing the new fields. The need to produce new hedges quickly encouraged exclusive use of the species *Crataegus monogyna*, or common hawthorn, a plant whose nickname, "quickset," attests to the speed and vigor with which it grew. The sudden explosion in demand for hawthorn drove the creation virtually overnight of a new nursery industry in northern England and Scotland.[48] In contrast to the ancient hedge embankments of Cornwall, with their diversity of plant and animal species, hedges in the parliamentary era were usually monocultures composed of two rows of hawthorn sets, one on each side of the property line. Finally, these new hedges and walls were often paralleled by new roads, up to thirty yards wide, designed to provide rapid access to expanding urban markets.[49] The result was a landscape of "sweeping sameness" marked by long walls, hawthorn hedges, and straight highways.[50]

Parliamentary enclosure was generally supported by both large and small landowners, but the requirement to fence proved difficult for individuals with limited access to capital and labor. While advocates of enclosure argued that fencing costs could be recovered as profit from land consolidation and conversion of arable land to pasture, many small owners were forced to sell off parts of their holdings to raise the money to construct hedges or walls on the remaining parcels. A parliamentary enclosure thus was often followed not by increasing diversity of ownership, but by consolidation of holdings in the hands of fewer and fewer people. As H. Rider Haggard demonstrated in his classic study of the parish of Feckenham in the Midland county of Worcestershire, in the late sixteenth century sixty-three owners had held title to 2,900 acres; by 1900, after enclosure, this number had fallen to six individuals holding 3,000 acres.[51] In many cases, then, parliamentary enclosure resulted in *fewer* fences delimiting *larger* holdings. By the early nineteenth century, the "single fence" that Thomas More had feared was well on its way to becoming a reality.

There is now little debate that enclosure was, in the words of historian Robert C. Allen, a "landlords' revolution" that transferred control over land and resources to an ever-smaller group of people.[52] Over the course of two centuries, smaller landowners such as yeomen were systematically eradicated from the landscape, in no small part by imposing fencing requirements that only the richest owners could meet. The effects of enclosure on small owners paled, however, in comparison with that vast majority of people who owned no land at all. Enclosures of open-field villages suddenly obliterated farming practices that were many centuries old. The hedges and walls of new parcels often crossed directly over the former planting strips, or "selions," that had ensured more or less equitable distribution of good and bad land among owners and tenants. The practice of grazing livestock in open fields had also ensured roughly equivalent amounts of fertilizer over the entire area of a village. After enclosure, owners and tenants cultivated crops or folded sheep on separate plots, whose quality, despite the putative efforts of the commissioners, varied greatly across the parish.

The second phase of parliamentary enclosure, of commons and "waste," was equally disruptive of ancient practice. As historian J. M. Neeson has convincingly shown, commons were critical and carefully managed resources in rural society. They provided a place to graze a single cow or sheep, as well as a reliable source of fuel, food, and building materials.[53] The removal of rights of access to these resources, rights that in fact predated those of any lord, was nothing short of calamitous for the poor and the landless.[54] In the words of one contemporary observer, commoners were quite simply "fenced out of their livelihood."[55] The common had made it possible to eke out an existence and to participate in "the network of exchange from which mutuality grew" even if one did not own land; enclosure removed that possibility for a large part of the rural population.[56] Even the requirements for new hedges were calculated to prevent their use by commoners, fruiting species such as *Mespilus germanica*, or common medlar, being discouraged because "the idle among the poor . . . would still be less inclined to work, if every hedge furnished the means of support."[57]

The disruptions of traditional practice caused by the fences and walls of parliamentary enclosure were famously depicted by John Clare in "The Moors," a poem that describes changes to Clare's own rural

21 "These paths are stopt": parliamentary enclosure hedges and selions of former open fields, Padbury, England, about 1950. Courtesy of Cambridge University Collection of Aerial Photography.

village, Helpston in Lincolnshire, written as he witnessed the enclosure of its open fields in the early decades of the nineteenth century.[58] The beginning of the poem paints a picture of an idyllic world where "unbounded freedom ruled the wandering scene / Nor fence of ownership crept in between." After enclosure, however, "fence now meets fence in owners' little bounds / Of field and meadow large as garden grounds / In little parcels little minds to please / With men and flocks imprisoned ill at ease." For Clare this new world of private parcels was far more than an aesthetic imposition. It fundamentally restricted ancient patterns of movement across the landscape:

These paths are stopt—the rude philistine's thrall
Is laid upon them and destroyed them all
Each little tyrant with his little sign
Shows where man claims earth glows no more divine
But paths to freedom and to childhood dear

A board sticks up to notice "no road here"
And on the tree with ivy overhung
The hated sign by vulgar taste is hung
As tho' the very birds should learn to know
When they go there they must no further go.[59]

Property holders are sovereigns in miniature, jealously guarding the bounds of their kingdoms, using hedges and fences as instruments for rescinding people's rights to the common, to say nothing of their most cherished memories. But landowners are worse than petty tyrants; in Clare's animist account, they are also "philistines," people of such "vulgar taste" that they can find it within themselves to nail "No Trespassing" signs to living trees. Clare thus draws the same parallel between fences and barbarism that Frost would use a century later in "Mending Wall."

In stark contrast to Young and other improvers, Clare saw in the fences and walls of parliamentary enclosure unmistakable symbols of a harsh new order where wage labor replaced self-sufficiency, and rights of property replaced rights of use. He was far from the last person to do so. Literary critic and novelist Raymond Williams wrote in the 1970s that "the many miles of new fences and walls were the formal declaration of where the power now lay. The economic system of landlord, tenant and laborer . . . was now in explicit and assertive control."[60] In these readings of history, by the early nineteenth century, boundaries in the English landscape had come to embody a potent dual sovereignty: that of landowners, on the one hand, and Parliament, on the other.

Never before parliamentary enclosure had walls and fences been drafted so quickly and consistently into the service of property. And yet, however radical the social and physical disruption they brought, enclosures by act of Parliament still took place in the context of landscapes dense with people and tradition. As the history of Otmoor shows, it was difficult to impose new boundaries without deliberation, opposition, and even violence. The enclosure commissioners were often forced to mold new enclosures to existing boundaries, if only to spare the landowners in whose interests they worked some of the cost of fencing new plots. Wresting property from "God's great common" often ran straight up against the social and physical realities of inhabited places.

The same cannot be said of England's newly independent colonies in North America. Here Locke's vision of ownership was being implemented on a scale that would in time come to dwarf the most ambitious commissioner's work—and every other project of enclosure before it.

Enlightened Allotment

As Cressey Dymock was drafting his plan to reclaim the wastes of East Anglia, on the other side of the Atlantic John Winthrop, the first governor of the Massachusetts Bay Colony, was making plans to claim the wilderness of New England. When English colonists landed in the New World in the 1620s, they had been met by Algonquian-speaking peoples who lived on cultivated maize and wild plants, and by hunting and fishing, and who moved their villages over the course of the year to exploit these resources.[61] Initial contact between the colonists and these peoples was peaceable; the Wampanoag, whose chief Massasoit had concluded a treaty with the Pilgrims at Plymouth, famously helped the new arrivals survive the harsh winter of 1620–21. With increasing migration in the mid-seventeenth century, however, the colonists began to press into the vast forests west of their original coastal plantations. The result was growing conflict with the indigenous population, culminating in King Philip's War in 1675–76. By the conclusion of that conflict, the Indians were decimated, their survivors fleeing west and north and opening New England to uncontested European settlement.[62]

Winthrop sought to construct a philosophical basis for this course of conquest. He did so in terms that would have been familiar to Dymock and Young. In his *Reasons for the Plantation in New England,* dated 1628, Winthrop justified the colonizing of Indian territory on the grounds that, unlike the colonists, the Indians "enclose no Land, neither have any settled habitation, nor any tame Cattle to improve the Land by . . . and so have no other but a natural right to those countries. So if we leave them sufficient for their own use, we may lawfully take the rest, there being more than enough for them and for us."[63] Here Winthrop was advancing, in rudimentary form, the same idea that Locke would fashion into a systematic theory of politics half a century later. Because the natives did nothing to fence their part of "God's great common," he suggested, they had forfeited all rights to inhabit and use it. This forfeiture paved the way for English colonial sovereignty.

Winthrop, who died in 1649, never questioned the status of New England as an English colony. Claiming the lands of North America was merely one part of the larger project of English imperial expansion, an essential component of building the English nation. But this was not Winthrop's only motivation. Like his correspondent Reverend Ezekiel Rogers, Winthrop was also a devout Puritan who advocated improvement on religious grounds. His desire to create a virtuous "city upon a hill" from the wilderness was so strong that Winthrop chose to remain in Massachusetts even when many Puritan settlers were returning to England to fight the Catholic Charles I in the English Civil War.[64] But more than religious fervor motivated Winthrop. The scion of a wealthy landowning family, Winthrop had seen his income from the rent of his 500-acre manor in Suffolk fall precipitously in the economic depression of the 1620s. His desire to enclose New England thus also had the status of a private commercial venture, one that would improve not only land, but also Winthrop's own finances.[65]

Winthrop's vision of North American colonization thus had three distinct components: political, religious, and economic. The act of enclosing land provided both the *justification* and the *instrument* of this tripartite sovereignty. Winthrop died long before he could witness the breaking away of the very colonies he saw as the vanguard of English imperial power. But whatever his status as a visionary, he could scarcely have foreseen the scale on which his project of national enclosure would be realized.

The United States was among the largest nations in the world when it was created, but it was also a country of ill-defined boundaries, "strained in coherence and constricted and unsatisfied in geographic position."[66] The need to subdivide large areas of land into new geopolitical units was a pressing problem from the very beginning of its existence.[67] Many of the country's founders sought to implement systems of territorial organization that would provide a political and spatial structure for the new polity. The early years of the Republic therefore saw a wide variety of experiments with rational forms of land allotment. James Oglethorpe's 1733 plan for the city of Savannah, an eighteen-square grid in the swamps of colonial Georgia, bore more than a passing similarity to Cressey Dymock's plan for the East Anglian fens a few decades earlier. And the New Military Tract, the system devised in 1782 to allot the wilderness of western New York to veterans of the Revolutionary War, resembled nothing so much as the Roman centuriations of Gaul and Iberia, an association reinforced by naming the square townships for Roman emperors, generals, and authors.

These early efforts were dwarfed by the subdivision of the Louisiana Territory, sold to the United States by Napoleon in 1803. The Louisiana Purchase, which doubled the new country's area overnight, was almost certainly illegal under the terms of the recently enacted Constitution, but several members of President Thomas Jefferson's cabinet justified it based on the United States' national interest in relation to the European powers.[68] In a letter to Jefferson dated January 10, 1803, Attorney General Levi Lincoln wrote, "The principles, and the precedent, of an independent purchase of territory, it will be said, may be extended to the east or west Indies, and that some future executive, will extend them, to the purchase of Louissiana, or still further south, & become the Executive of the United State of North & South America."[69] In other words, the Louisiana Purchase was the first step in an inevitable teleology that would lead the United States to colonize the entire area west to the Pacific Ocean and south to the Straits of Magellan. Jefferson eventually overcame his reservations.

Like the other men who laid the political foundations of the new United States, Jefferson was a child of the British and French Enlightenment, and his ideas were deeply influenced by Locke's thought on property, the social contract, and the nature of representative government. Like Locke, Jefferson believed that governments served at the pleasure of landowners who held the franchise; both men were essentially Whig aristocrats who believed that the main function of the social contract was to protect individuals' right to property. But in sharp contrast to Locke, Jefferson believed that the basis of the American republic lay in its future as an agrarian state in which all (male) citizens owned land.[70] It was not by expanding the franchise to non-landowners that the United States would become more democratic than its European counterparts, in other words, but by making every adult male a landowner.[71] The political destiny of the United States as Jefferson saw it was to implement Locke's vision of property on a scale more extensive than Locke himself ever foresaw. The acquisition of Louisiana provided the opportunity to give spatial expression to this political and economic vision. The result was the largest single program of land allotment in history.

The roots of this program went back nearly two decades, to the day in 1784 immediately following the ratification of Virginia's claim to the land northwest of the Ohio River. On that day the secretary of Congress appointed a five-member committee, chaired by Jefferson, to "devise and report the most eligible means of disposing of such part of the Western lands as may be obtained of the Indians."[72] The Francophile Jefferson was convinced of the superiority of the new French decimal system from his experience designing the national currency, and he proposed dividing the land west of the Appalachians into a grid of "hundreds," or ten-mile-square townships each subdivided into one hundred lots of 850 acres. What resulted, after much deliberation, were the "Northwest

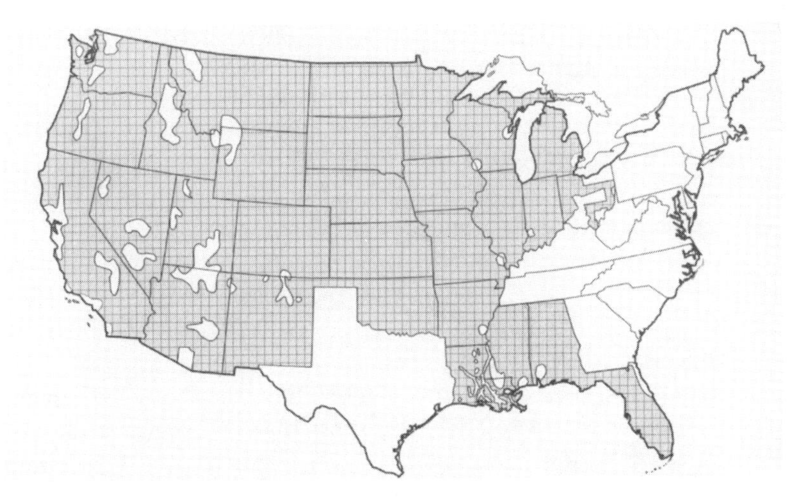

22 Sovereignty by allotment: lands subdivided under the United States Northwest Ordinances of 1784, 1785, and 1787. Courtesy of Lawrence Vale/Harvard University Press.

Ordinances" of 1784, 1785, and 1787. The Land Ordinance of 1784 (later replaced by the 1787 Ordinance) organized Jefferson's townships into a series of self-governing districts, each eligible to send one representative to Congress when it had attained a population of 20,000.[73] The Land Ordinance of 1785 specified the method of surveying the townships. It retained the principle of square divisions oriented along the cardinal compass points but replaced Jefferson's decimal hundreds with a duodemical system of thirty-six-square-mile townships subdivided into sections of one square mile, or 640 acres.[74] The section numbered sixteen, near the center of each township, was reserved for enterprises generating income for a public school, but the location of everything else was left to future settlers.[75]

The impulse behind the Northwest Ordinances was far from new; land has been subdivided and allotted since the ancient Egyptians. But this should not obscure the radical nature of the project the American legislators envisioned. The most distinctive feature of the ordinances was the absolute inflexibility with which they imposed a wholly abstract "order upon the land."[76] Compasses had been available since the twelfth century and used for surveying and laying out fields at least since the sixteenth.[77] But it was one thing to use a compass to bound a single enclosure and quite another to use it to bound an entire continent. "The old reliance on natural edges and shapes passed" as an abstract two-dimensional geometry was forced onto a sphere by staggering north-south boundaries at regular intervals.[78] This was a significant departure from older subdivisions, even in North America.[79] An allotment in North Carolina completed less than a decade before the Northwest Ordinances, while describing tracts of "right Lines, running East, West, North and South," had nevertheless allowed for "the use of natural boundaries as an alternative way of establishing readily verifiable limits," and it exempted parcels adjacent to older land grants or navigable waterways from subjection to its orthogonal logic.[80] The Northwest Ordinances made no such concessions. Maps showed lines of allotment running over mountains, rivers, gullies, and streams (and, implicitly, the people who inhabited the areas), a spatial representation of the political, economic, and cultural sovereignty of the new country.

Despite their power as a graphic symbol, however, the Northwest Ordinances were more speculative than substantive. Drawing a grid of allotment across an entire continent was easy; giving the resulting boundaries material shape on the land was not. Yet Locke, whose thought underlay the entire project, had argued that physical enclosure was the first act of property and hence the social contract; Jefferson himself wrote at length of his own wooden fences at Monticello, calling them "a great & perishable work" and taking considerable pride in the ten-foot-high palings around his kitchen garden.[81] The ordinances thus harbored from the very beginning a fundamental tension. If lots could not be marked by actual constructions, the sovereignty they represented would forever remain in doubt. In the decades of western settlement that followed passage of the ordinances, this tension would become more acute than Jefferson, sitting on his mountaintop estate rich in timber and stone, could have imagined.

Jefferson and other Republicans foresaw settlement beyond the Appalachians as an orderly process in which indigenous populations—the "non-enclosers" described by Winthrop—would give way to a democratic republic of farmer-owners.[82] In this vision the surveyors were the advance guard of the settlers. Throughout the last decade of the eighteenth century, numerous expeditions were sent to survey and map the lands of the West before migrants arrived. The most famous of these expeditions was that led by Meriwether Lewis and William Clark, dispatched by Jefferson in 1803 to map the territory acquired through the Louisiana Purchase, a vast area corresponding to the modern states of Missouri, Kansas, Nebraska, South and North Dakota, Montana, Idaho, Washington, and Oregon. By staking out the boundaries of the square townships prescribed by the Land Commission, Lewis and Clark and the legions of surveyors that would follow would create ready-made receptacles for the agrarian communities Jefferson and other Republicans envisioned.

Events rarely worked out this way. Surveyors were often among the *last* to arrive, anticipated by squatters, farmers, ranchers, and prospectors.[83] The reality on the ground, by the mid-nineteenth century, was a dense patchwork of squatters' claims alternating with large swaths of territory controlled, though often not legally owned, by eastern cattle interests and railroads.[84] Whatever their economic and social difference, these migrants were united by their indifference to the philosophical speculations of Locke and Jefferson. The orthogonal section boundaries of the Land Ordinances were widely ignored, and foot traffic and cattle meandered among irregular holdings largely without regard to statutory divisions.[85] It was a landscape shaped by common need rather than the compass needle.

The political debate about the legitimacy of these migrants divided the United States Congress in the early nineteenth century. On the one hand, many Whigs from the East and South described squatters in terms similar to those Arthur Young was using in England to describe commoners. Squatters were "lawless vagabonds" whose unsanctioned presence on land they did not own threatened to undermine the foundation of stable communities. "Antisquatter congressmen feared that if squatting continued, the West would become a region of thinly scattered barbarians who took their livings off the richest lands but who were unable to support the basic institutions of the republic."[86] On the other side of the debate, westerners and most Democrats extolled squatters as the vanguard of Jefferson's agrarian republic. Here, supporters argued, were Locke's true enclosers, wresting their part of "God's great common" by dint of sheer industry. Thomas Hart Benton, among the most forceful advocates for squatting, hailed "the people going forward without government aid or countenance, establishing their possession and compelling the Government to follow with its shield and spread it over them."[87]

Such sentiments won the day, and in 1841 a "permanent preemption act" was passed guaranteeing the first squatter purchase rights to the section in which his improvements lay.[88] These rights were then greatly expanded in the Homestead Act, signed by Abraham Lincoln in 1862. The act, which covered nearly all the land of the Great Plains and the West, granted any settlers free title to a 160-acre parcel of land (one-quarter of the 640-acre sections originally laid out by the 1785 ordinance, hence the name "quarter section") if they inhabited and improved it for five years. Never before had Locke's vision of property, as the mixing of labor with the common, been given this kind of statutory force. The attraction of the scheme was irresistible to many migrants, who flooded into the lands west of the Missouri River in the decades after the act's passage. But the Homestead Act was far from the only means the American government was using to carve up and distribute land. Railroads had already obtained title to huge swaths of territory, and land speculators were rapidly consolidating holdings by acquiring grants from their original holders at below-market rates.[89] Many homesteaders thus arrived only to find most of "God's great common" already spoken for. "A settler entering Kansas in the late 1860s and early 1870s, for example, would find one-third of that state closed to homesteading. Railroad grants alone tied up 20 percent of the state."[90]

In the struggle to secure title to a claim, the small homesteader was at a distinct disadvantage. Most of the areas covered by the Homestead Act, from endless grasslands to rugged mountains, simply did not provide the raw material to enclose land on the vast scale envisioned by American legislators. In the arid lands of the Great Plains and the West, stone and timber were often almost nonexistent; the chief means of demonstrating "improvement"—building fences—was thus unavailable.[91] Even in those places where materials were present, the sheer size of the quarter section, whose perimeter was two miles, made fencing on any large scale virtually impossible.[92] The inability to physically enclose land, along with inappropriate climatic and soil conditions, was thus a principal reason many homesteaders were ultimately unable to gain title to their claims.[93] Even in a state with excellent agricultural conditions, such as Kansas, only 41 percent of settlers filing a claim on a quarter section between 1862 and 1873 managed to "prove up" that claim; for settlement under the Homestead Act as a whole, the figure was only 33 percent.[94]

Ironically, securing claim was easier for the absentee owners and land speculators, often with virtually unlimited capital, who were rapidly gaining control of entire regions by acquiring adjacent parcels.[95] In the years following the passage of the Homestead Act and, later, the Timber Culture Act of 1873 and the Desert Land Act of 1876, the West was increasingly dotted with all kinds of provisional objects establishing claims to land. Such transparently fraudulent "improvements," from irrigation ditches that went nowhere to tool sheds miles from the nearest town or railroad, were frequently condemned, but this did not stop them from meeting the letter of the law of improvement.[96]

Whatever their differences, speculators and homesteaders were thus presented with a common problem: how to *enclose* the common and create property where the material for physical enclosure was scarce. By the end of the 1870s, however, both large and small landowners would gain access to a cheap and virtually inexhaustible device for asserting their respective forms of sovereignty over the territory of the interior United States. Ironically, the material that would finally solve "the fencing problem" nearly a century after Jefferson conceived his land allotment system was the product of the very mechanization he had viewed as threatening his ideal agrarian republic. And indeed, the way this material was used would come to resemble not the social contract of Locke, but the state of nature of Hobbes.

In the years immediately following passage of the Homestead Act, many settlers had responded to the chronic shortage of fencing materials on the Plains by improvising fences out of steel wire. But wire was expensive and generally unavailable in the vast quantities necessary to fence a quarter section. Where they were built, wire fences tended to sag in summer and break in winter. The 1860s therefore saw a number of attempts to improve the durability of wire fencing, culminating in 1867 in the first patents for wire with barbs or "thorns," taken out within a month of each other by William D. Hunt in New York and Gilbert Gavillard in France.[97] These early designs suffered from an important problem, however: the tendency of the barb to slip on the wire. In 1874 J. F. Glidden of DeKalb, Illinois, patented a modification of Hunt's invention in which two wires were twisted together and the barbs were fixed firmly between them. Glidden sold a half interest to a local hardware dealer, Isaac Ellwood, and the two began producing several thousand pounds of the material every year.

"Barbed wire" was exceedingly easy to manufacture. Given a small amount of capital, almost anyone could start a factory producing it.[98] This simplicity drove rapid growth in production during the last decades of the nineteenth century, primarily along the rail lines of the upper Midwest. By 1883 there were at least thirteen barbed-wire manufacturers in the vicinity of DeKalb; Chicago and Joliet had eight factories apiece in the same year, and St. Louis had eleven factories three years later.[99] Production exploded from 80,000 tons between 1880 and 1884, to over 157,000 tons in 1895 alone. The result was a drastic fall in price. Barbed wire had cost a prohibitive twenty cents a pound when production began in the 1870s; by 1893 the price had fallen to two cents.[100]

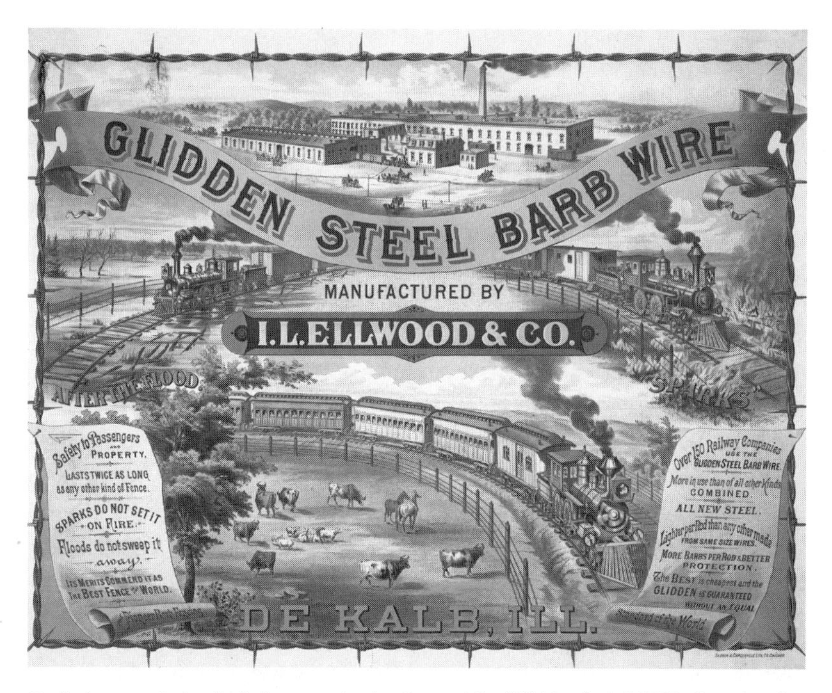

23 Enclosure as technological conquest: advertisement for "Glidden Barb," 1877. Courtesy of the Baker Library, Harvard University.

Unlike stone walls and wooden fences, barbed wire could be installed over large areas quickly and cheaply and required little maintenance once in place. Small farmers moving into the West under the Homestead Act thus gained a vital new tool for demonstrating improvement and gaining title to land. By the 1890s, a settler could enclose a field with a three-wire fence for approximately $150 an acre, or less than half the cost of boards and pickets.[101] Barbed wire and wood posts were rapidly distributed by the new rail networks fanning out across the Plains and were often sold directly off train cars. As migration under the Homestead Act increased and the standard allotment size was reduced, demand for wire exploded, since the smaller the enclosure, the greater the length of fence per acre.[102] Within a few years, then, it had become possible to fence vast areas at low cost and with minimal effort. The result of this ease of fencing was hardly the orderly process of settlement envisioned by Jefferson or Benton, however. Instead, the sudden availability of cheap and plentiful fencing brought the battle for sovereignty over the West to a new level.

At the center of this battle was a fundamental conflict over how land would be used and who would control it. Homesteaders were seen by the federal government as the ultimate highest users, since it was they who would transform the West into a land of freehold farmers, the equivalents of England's yeomen. But this notion often ran aground on the shoals of climatic reality. A single 160-acre quarter section was not nearly enough to support a viable farm. In many parts of the West, therefore, agriculture had only the most tenuous of holds. In its place large cattle interests, controlled from the financial centers of the East or Europe, had moved in to colonize enormous areas of open range. As more and more homesteaders arrived in the 1870s and 1880s and began to claim even marginal areas, a desperate battle took place between these two very different users of the land.

Barbed wire was the main weapon in this battle. Many cattle interests and ranchers held no legal title to the land they used, but the availability of virtually infinite amounts of fencing meant that suddenly they could give material shape to their claims, enclosing vast areas of range for the exclusive use of their herds.[103] Legal boundaries meant little as barbed wire was "thrown up everywhere, irrespective of titles, roads, or laws."[104] "It is doubtful if the world has ever witnessed such criminal prodigality," the commissioner of the General Land Office wrote of the cattle barons. "Whole counties have been fenced in by the cattle companies, native and foreign, and the frauds that have been carried on by individuals on a small scale are simply innumerable."[105] These events were captured by Mollie E. Moore Davis in her novel *The Wire Cutters*, published in 1899:

> Gangs of laborers were at work putting down cedar posts, and stretching interminable shining lines of barbed wire in every direction. There was a certain recklessness in the proceeding, a disregard both of possible ownership and of propriety, which excited profound indignation. In some quarters the public roads and familiar byways were closed up; many small freeholds, belonging to the poorer men, were fenced in; even larger places were practically barred from communication with the common highways. But worst of all, the water supply, scarce at all times in this region, was in many instances cut off; springs, ponds, and water-holes were ringed about with the formidable wire.[106]

This passage suggests how greatly barbed wire changed patterns of human use and movement. Smaller farmers and ranchers had seen unfenced land as a kind of common, a public resource capable of supplementing the quarter section plots to which they had legal claim. Before the introduction of barbed wire, homesteaders and squatters had tended to ignore the surveyed paths and roads of the square land divisions, taking shortcuts across fields. Since barbed wire could extend for many miles without a gate, it suddenly forced residents to travel along roads, vastly increasing travel times.[107] Yet barbed wire *itself* did not always follow roads. One settler recalled fences "that stretched for miles across the plains irrespective of roads or trails," threatening "the cheering influences of Church & School" by preventing "further settlement of the public domain."[108] Many highways were reduced to little more than cow paths as using them required travelers to go through scores of gates; barbed wire even disrupted the nascent mail service.[109] And during the drought of the early 1880s, desperate homesteaders were confronted with the fact that water holes once open to all were now fenced in.[110] Never had Thomas More's "single fence" come closer to landscape reality.

The result of these disruptions was an explosion of "fence wars" between ranchers and smallholders in the 1880s, with ranchers destroying the enclosures of homesteaders and homesteaders destroying the (far larger) enclosures of ranchers.[111] In Texas, the fence-cutting war of 1883–84 resulted in numerous deaths in Clay and Brown Counties; the state legislature responded with a law declaring fence cutting a felony, punishable by up to five years in prison.[112] In San Miguel County, New Mexico, a "bitter dispute developed in 1889 between Hispanic villagers who attempted to defend their communal grazing lands from Anglo cattle ranchers and merchants who claimed that the lands were not communal but were partible—that is, they could be divided and sold. When ranchers fenced the land, masked and armed horsemen . . . destroyed the fences."[113] And in the most famous incident, small ranchers in Johnson County, Wyoming, who had contested "illegal monopolists of the public domain" were attacked by a group of fifty Texas gunmen and "regulators." The vigilantes were held at bay by a group of two hundred settlers; remarkably, only one person died in the skirmish.[114]

These confrontations quickly became the stuff of popular myth. As reported by the unsympathetic *Galveston Daily News* in 1885, wire cutting "found its way to the fireside of every home, [where] the grievances of the lawless element of the communistic fence-cutters were held up in glowing colors."[115] The writer's emphasis on the "communistic" overtones of fence cutting was hardly incidental. Smallholders were profoundly aware that by cutting ranchers' fences they were asserting their rights to those commonly held resources, whether land or water, that their survival depended on. They saw barbed wire as the embodi-

24 Otmoor revisited: barbed-wire fence cutters in Nebraska, 1885. Courtesy of the Nebraska State Historical Society.

ment of an illegal annexation of the public domain, a "system of spoliation" that threatened to undermine the formation of stable agricultural communities.[116]

The threat of barbed wire to homesteaders paled, however, in comparison with its effect on the people who "enclose no land." In the early years of its existence, barbed wire had been banned in many states for its perceived cruelty to domestic livestock.[117] Yet this very tendency made it useful for the one project homesteaders and ranchers could agree on: the eradication of the western bison and the Indians whose economy depended on it. Throughout the 1870s, increasingly dense mazes of wire in the western landscape trapped bison, making them easy targets for travelers, who picked them off from passing trains; it is estimated that 5.5 million bison were killed this way in the 1870s alone.[118] Barbed wire also blocked Indians' access to traditional hunting grounds, making their "geographical and social environment hostile to them, so that it became a foreign territory where the tribal way of life was unimaginable."[119] However barbed wire may have threatened the emergence of Jefferson's agrarian communities, in other words, it continued a teleology of national expansion and sovereignty over the Indians that reached back to John Winthrop.

The invention of barbed wire was a seminal event in the history of the modern landscape. Its low cost, ease of installation and maintenance, and effectiveness at controlling livestock made it a kind of universal tool of enclosure, giving European settlers the means to gain a foothold in a hostile country and demonstrate the Lockean improvement that would one day make them property owners "equal to the greatest." But these same characteristics also made barbed wire indispensable for cattle interests, railroads, and land speculators. Even as barbed wire "aided [the] small farmer to gain a foothold in the Great Plains; it enabled the cattleman as well to secure and hold range land for his herds."[120] However useful to "actual settlers," barbed wire was no less useful for asserting control from afar, even when that control had no basis in law.[121] It was thus not long before barbed wire was seen by many as symbolizing not improvement by new citizens, but brazen theft of common resources. Barbed wire may not have been used by rural aristocrats asserting sovereignty over their estates, as were the hedges of parliamentary enclosure. But as it began to be used in the decades after its invention, it was the tool of a bigger, more elusive, and ultimately more powerful sovereign: capital, slowly but inexorably consolidating its hold on the western frontier.

It is difficult to overstate how greatly barbed wire transformed ancient practices of marking land. No longer was it necessary to plan and negotiate boundaries, even as had been done during the parliamentary enclosure era in England. In earlier times, the very difficulty of producing boundaries as objects had made it virtually impossible to build them without accountability and some degree of deliberation; constructing any fence necessarily involved talking to people. With barbed wire, erecting a fence could take days or even hours rather than months. What had once required the effort of many people could now be accomplished by one or two. It was now possible to enclose land cheaply, anonymously, and virtually overnight, characteristics that were to make barbed wire the preferred material of social control and warfare in the twentieth century.[122] Barbed wire was also the first truly modern fencing material in the method of its production and distribution. No longer did fences have to be constructed from materials close at hand; barbed wire could be produced in one place for use hundreds of miles away, acquired through a simple market transaction from the passing trains whose owners were among its biggest consumers.[123]

But the most striking feature of barbed wire was its ability to assert control over territory while leaving only the slightest trace on the land. Barbed wire thus not only was a means of asserting claims of sovereignty; it was also a way to disguise, mystify, and naturalize those same claims. This capacity to combine *control* with *disappearance* was a crucial innovation, one that would come to shape the ways boundaries were marked thenceforth, not just in America's great arid expanses, but on the margins of its burgeoning cities.

Disappearing the Wall

As the landscape of Middle England was being covered by walls and fences in the late eighteenth century, a new aesthetics of property began to emerge among the class whose interests they served. The new field boundaries were clear and unmistakable expressions of landowners' newly consolidated sovereignty. They were not just practical devices for enclosing livestock, but pervasive symbols of the economic and social control of one group of people by another. But in the pleasure gardens that abutted their houses, landowners were rapidly embracing a new and radical style, one that recalled nothing so much as the very open field landscape lately eradicated by their own actions.

One of the earliest examples of this style was Blenheim Palace, near Woodstock in Oxfordshire, where in 1764 the landscape gardener Lancelot "Capability" Brown (1716–83) was asked to improve the grounds originally laid out by the architect of the mansion, Sir John Vanbrugh. Brown replaced Vanbrugh's gravel parterre with an enormous greensward that swept down from the mansion, with clumps of trees and shrubs artfully arranged along its margins. At the center of the composition was a new serpentine lake formed by damming three streams. The effect was what Uvedale Price, the advocate of the "picturesque" style, called "a very just and easy gradation from architectural ornaments, to the natural woods, thickets, and pastures."[124]

Brown's design was a decisive turn away from the aesthetics of the Baroque, with its straight lines and hard edges, toward the flowing, naturalistic style of Romanticism. Yet even before he set to work, Brown was significantly aided by an innovation in Vanbrugh's original design. This innovation lay at the very margins of the mansion grounds, in the way the formal garden met the surrounding fields. Rather than proposing a fence or hedge, which would have delineated the area immediately around any large country house in previous centuries, Vanbrugh had enclosed the garden with a "ha-ha," a shallow ditch used for controlling stock where timber and stone were scarce. As described by Horace Walpole, the omnipresent arbiter of taste, the "sunk fence" enabled the "contiguous ground of the park . . . to be harmonized with the lawn within; and the garden in its turn . . . to be set free from its prim regularity, that it might assort with the wilder country without. The sunk fence ascertained the specific garden, but that it might not draw too obvious a line of distinction between the neat and the rude, the contiguous outlying parts came to be included in a kind of general design."[125]

The ha-ha kept sheep out while visually fusing the surrounding landscape with the formal garden. But this new device also had a symbolic resonance that was unmistakable in its social context. Concealing the boundary made the great house visible from all the lands around it, one of many "strong points of a class" rising in the landscape, and gave its inhabitants a clear view of the territory over which they held almost complete sway. Like the new fences and walls farther afield, the exposed mansion was a clear embodiment of landowners' economic and political dominance, a "visible stamping of power, of displayed wealth and command: a social disproportion which was meant to impress and overawe."[126] Control of land and people was thus suddenly expressed not only or even primarily by the presence of walls, but by their absence.

The redesign of the gardens at Blenheim was merely an early stage in the "imparkment" of the landscape that would follow the enclosures of the late eighteenth century. During this period the landlord class, flush with the new income provided by converting arable land to pasture, transformed their country houses into mansions surrounded by "landscape parks" with the same sweeping lawns, the same serpentines, the same invisible walls. Whether at Rousham, Stowe, or Heveningham, the result was the same: as the newly enclosed fields belonging to the estate "began to look increasingly artificial, like a garden, the garden began to look increasingly natural, like the pre-enclosed landscape."[127] The implications of this change would be decisive for the aesthetics of bounding in the centuries that followed.

If Lancelot Brown could lay claim to the most aristocratic clients, the honor for most commercially successful designer must go to his aesthetic heir, Humphry Repton (1752–1818). Repton was born to a wealthy family that had marked him for a career as a merchant. He failed to achieve success in this area, however, and chose instead to retire to the country and develop his skills as an amateur watercolorist. In 1788, at what was then the advanced age of thirty-six, Repton decided to establish himself as a landscape designer to the gentry, soliciting financial support from a list of wealthy friends that included the Duke of Portland and Thomas Coke, First Earl of Leicester, one of the principal agricultural improvers of the time.[128]

Like Brown, Repton was a strong advocate of obliterating physical boundaries between the country house, its garden, and the wider landscape. The main justification for doing so, he argued, was to open the view both *from* and *toward* the mansion. In *Fragments on the Theory and Practice of Landscape Gardening* (1816), he instructed his readers that "if there be any part of my practice liable to the accusation of often advising the same thing at different places, it will be true in all that relates to my partiality for a *Terrace* as a fence near the house."[129] Repton's so-called Red Books employed his watercolor skills and an ingenious system of sliding panels to show clients what they would see from their windows, and what visitors would see on the drive up to the house, once he had been given leave to remove or conceal unsightly fences, walls, and hedges. In the Red Book for Glenham Hall, for example, Repton showed before-and-after views of the grounds from a window of the mansion. The first image depicted a view of a bare hillside partially blocked by an enclosing hedge around the property; in the second image, the hedge had been replaced by a sweeping serpentine and the wooded slope beyond.[130]

Repton's objection to physical boundaries was more than aesthetic; it also had a moral and economic dimension. Repton believed strongly in the superiority of the gentry, and he saw the merchants and bankers who were buying up country houses in the mid-nineteenth century as interlopers and parvenus, too intent on overt display of their growing control over the landscape. These owners' tendency to enclose their new estates with high fences and walls was for Repton among their most grievous transgressions. In one passage from the *Fragments*, Repton drew alternative views of the edge of an estate where it met a public road. The first showed a modest rail fence that nearly disappeared in a picturesque scene, giving a clear view into the property; the second showed the high palings that had replaced it, presumably erected by a "worthy cockney" who had bought the property from an impecunious earl. The pair of images was ironically labeled "Improvements."[131]

Despite his disdain for them, these worthy cockneys were more and more often Repton's main clients. If Repton had begun by designing landscapes for the rural aristocracy, he ended his career laying out the far more modest grounds of the growing middle class, men who would engage him for a single afternoon to "stake out the ground and site for a villa" in the rapidly growing suburbs of London.[132] Repton acknowledged the "unpardonable extravagance in making large plantations" in environments dense with houses.[133] But designing gardens in suburbs brought with it a distinct problem: how to achieve the illusion of openness and amplitude, the primary characteristics of the rural landscape park, in a landscape increasingly subdivided into small lots. Repton's design for his own garden showed in a particularly revealing way just how he proposed to solve that problem.

The merchants and bankers Repton worked for not only were his clients—they were also his neighbors. For most of his adult life, he lived in a small cottage at Hare Street in Essex, on the edge of London's expanding suburbs.[134] Repton's garden fronted the village common, a triangle of ground at the intersection of two busy roads. The position was advantageous for its ease of travel, but Repton decried the state of the common and its visual relation to his garden. "A view into a square, or into the Parks, may be cheerful and beautiful," he wrote, "but it wants appropriation, it wants that charm which only belongs to ownership; the *exclusive right* of enjoyment, with the power of refusing that others should share our enjoyment."[135] In 1802, after many years of petitioning, Repton finally received permission to undertake such appropriation around his own property.

The success of this petition was no doubt in part a function of Repton's visual rhetoric. In the *Polite Repository* of 1800, Repton published before-and-after drawings similar to those he had perfected in the Red Books. In the first image, the garden faced directly onto the common; across the front of the image was a fence of wood lattice; a beggar, hat in hand, hung over the railing. Behind the beggar, the common was populated by little more than a gaggle of geese (the carcasses of several of their fellows were hanging on a porch opposite), and coaches were visible on the main road. Two bare tree trunks stood at either edge of the scene. In the second image, all hint of hard edges had disappeared. The fence had been replaced by a new lawn sweeping out to a hedge and orchard; the goose carcasses were concealed by a trellis, and the nearest coach and beggar were gone. Flower beds framed a waiting chair and watering can; a hydrangea climbed the trunk of one of the trees.[136] "I have obtained a frame to my Landscape," Repton boasted.[137] The scene was a Glenham Hall in miniature, created and maintained by Repton's own labor.

But the most striking part of the second image was the *location* of

25 Enclosing the common, obscuring the fence: Humphry Repton's garden expansion, Essex, England, about 1800. Courtesy of Yale Center for British Art.

the boundary between parcel and common. In the first image, the fence ran directly across the foreground; the beggar was so close one could almost make out his teeth. A public road ran on the far side of the fence along one edge of the common. In the second image, the new hedge ran boldly across the middle ground, and the public road had become a path in the garden. The former common, in other words, had been entirely absorbed into Repton's own yard.

Repton thus demonstrated, on his own doorstep, how the landscape park could be reconciled with the spatial limitations of small parcels (and the economic limitations of their owners). But the achievement that Repton's garden represented was as symbolic as it was practical. Eliminating all obvious markers of the boundary of the parcel made a political and economic act—the enclosure of a real common and its alienation into private ownership—appear "natural" in the same way that the ha-ha had served to naturalize, in the rural landscape of the eighteenth century, a particular set of social and economic relationships between those who owned property and those who did not. This association between the aesthetics of openness and the reality of property would soon be brought to greater and greater parts of the suburban landscape in England—and then, with an abandon that likely would have surprised Repton himself, America.

One man who took up this association after Repton's death was the Scot John Claudius Loudon, born in 1783. Loudon was less interested in the gardens of the rural gentry than in Repton's later clientele, the emerging middle class of suburban landowners then buying up properties on the edges of Manchester and London.[138] Loudon conceived gardens as settings for suburban dwellings that would replicate, in aesthetic sensibility if not in scale, the great houses of rural England. In 1803 he moved to London, where he established himself as a landscape architect, a term he would coin in 1840.[139] Loudon's book *The Suburban Gardener,* first published in 1838, attained great commercial success in the first decades of the nineteenth century. In the book Loudon advocated a style he dubbed "gardenesque," a reference to the "picturesque" aesthetic advanced by Uvedale Price and Richard Payne Knight in the 1790s.[140] Like the picturesque, the gardenesque departed from the sweeping lawns and tree clumps of Brown and Repton to emphasize "natural" irregularity, crookedness, and decay.

But in one crucial area, Loudon was in agreement with his predecessor. Like Repton, Loudon saw overt parcel boundaries as coarse reminders of property and status, to be avoided if at all possible. In one passage of *The Suburban Gardener*, on the garden of Hendon Rectory, Loudon lauded the designer for "an irregular hedge with oaks and elms, which harmonizes so well with the adjoining fields similarly enclosed, that the limits of the property are in no way discernible."[141] Similarly, Loudon recommended placing "a column surmounted by a statue, or an obelisk . . . to divert attention from the boundary fence."[142] Walls were permissible but should be no more than three feet high, "sufficient to keep out vermin," and in most cases a more subtle "holly hedge" was preferable to a wall.[143] Like ostentatious houses, fences and walls were the marks of money recently gotten; real wealth, economic and cultural, was confident enough of its prerogatives to blend its gardens with the landscape as a whole.

Shortly after its publication, *The Suburban Gardener* fell into the hands of an ambitious young American, Andrew Jackson Downing. Born in the Hudson River town of Newburgh, New York, in 1815, Downing had attended an exclusive boarding school where he learned to draw, but this was the extent of his formal education. Upon the death of his father, Downing took over the family nursery, and in 1838 he married into the family of a wealthy land speculator with investments in railways and ferry lines. Using capital from this alliance, Downing built a Gothic Revival house for himself and his wife on a parcel inherited from his parents. From this location Downing embarked on a brief but astonishingly productive career as a "horticulturist" and arbiter of taste for the emerging industrial elite of the United States, publishing numerous treatises and pamphlets on gardening and domestic architecture until his untimely death in a steamboat fire in 1852.[144]

The first and most important of these publications, the 1841 *Treatise on the Theory and Practice of Landscape Gardening, Adapted to North America*, was a rendition of *The Suburban Gardener* for an American audience. A tireless promoter of "country places" for the well-to-do, Downing considered the winding lanes and turreted cottages of Loudon's "gardenesque" style ideally suited to the topographical and horticultural conditions of the Hudson River valley, where merchants and industrialists from New York had begun to build part-time residences for themselves and their families. Despite its provenance, Downing argued, the gardenesque was at its root an *American* aesthetic, one that, like the paintings of the Hudson River school, contributed to the larger project of building the new nation.[145] "While I have availed myself of the works of European authors, and especially those of Britain, where landscape gardening was first raised to the rank of a fine art," Downing wrote, "I have also endeavored to adapt my suggestions especially to this country and to the peculiar wants of its inhabitants."[146]

This vision of a national aesthetic was paired with a moral vision based on the nuclear family and private home. In his books and in the *Horticulturist*, the journal he edited from 1845 until his death, Downing tirelessly promoted the idea that the "simple cottage" and garden were the prime repositories of moral value. "The family, whose religion lies away from its threshold," he wrote, "will show but slender results from the best teachings, compared with another where the family hearth is made the central point of the Beautiful and the Good."[147] The prime object of landscape gardening was to "embody our ideal of a rural home; not through plots of fruit trees, and beds of choice flowers . . . but by collecting and combining beautiful forms in trees, surfaces of ground, buildings, and walks, in the landscape surrounding us."[148]

Such love of home was closely connected to love of country at a time of rapid change in American society.[149] Downing saw the "rural residence" as a refuge from the urbanization and industrialization that were rapidly transforming the American landscape and American culture. Though Downing recognized that it was "needful in civilized life for men to live in cities," he held that "in the United States, nature and domestic life are better than society and the manners of towns. Hence all sensible men gladly escape, earlier or later, and partially or wholly, from the turmoil of cities."[150] Downing fervently opposed the "mere rows of houses upon streets crossing each other at right angles" that characterized most American cities and condemned the speculative developer who "covers the ground with narrow cells, and advertises to sell or rent them as charming rural residences."[151] This contempt for the urban gridiron reflected a wider distaste for geometrical form, which Downing explicitly equated with the iniquity and tyranny of the Old World.[152]

Downing's antipathy toward urbanization was profoundly ironic. If the city was a demonic machine that "all sensible men gladly escape," it was also, as it had been for Repton and Loudon, the source of his own clients. The "simple cottages" described in the *Treatise* and the *Horticulturist* were far beyond the means of ordinary people, and they were routinely mocked in the agricultural press for their cost and inefficient layout.[153] Downing's primary audience was urban, not rural; the "country places" he promoted differed little, in terms of the economic forces driving their construction, from the speculative "narrow cells" being laid out closer to the old city centers.

There was thus a tension in Downing's moral and aesthetic vision, based on year-round country living, and the economic and social reality of expanding cities. The main problem Downing faced was therefore similar to that confronted by his predecessors: how to sustain the illusion of rural amplitude in an ever-denser suburban landscape. This required considerable skill on the part of the horticulturist:

> Suburban villa residences are, every day, becoming more numerous; and in laying out the grounds around them, and disposing the sylvan features, there is often more ingenuity, and as much taste required, as in treating a country residence of several hundred acres. In the small area of from one half an acre to ten or twelve acres, surrounding often a villa of the first class, it is desirable to assemble many of the same features, and as much as possible of the enjoyment, which are to be found in a large and elegant estate.[154]

As in England, the desire to render the suburban parcel an estate in miniature had profound implications for the treatment of boundaries. Like Loudon and Repton, Downing advocated concealing neighboring properties and public streets with trees and shrubs in order to "open to the eye, from the windows or front of the house, a wide surface, partially broken up and divided by groups and masses of trees. . . . In more distant parts of the plantations will also appear masses of considerable extent, perhaps upon the boundary line, perhaps in particular situations on the sides."[155] Such disposition, "where no boundaries are conspicuous, conveys an impression of ample extent and space for enjoyment."[156] Downing also recommended ample planting to disguise the new systems, primarily roads, that linked the "country place" to the city it depended on. In an 1850 editorial, he commanded his readers, "Thou shalt plant trees, to hide the nakedness of the streets."[157] The roads and cheek-by-jowl parcels of new suburbs were unseemly reminders of the creeping city just over the horizon, and they should be concealed by any means possible.

This idea explains the almost feverish antipathy with which Downing viewed fences, which he called, presaging Frost's pun, "among the most unsightly and offensive objects in our country seats."[158] "To fence off a small plot around a fine house," he wrote, "is a perversity which we could never reconcile, even with the lowest perception of beauty. An old stone wall covered with creepers and climbing plants, may become a picturesque barrier a thousand times superior to such a fence. But there is never one instance in a thousand where any barrier is necessary."[159] Downing recommended that any fences near the house be hidden by "plantations."[160] Of course, such revulsion was possible only because the new suburban property owner did not face the same practical requirements as real farmers did, particularly fencing livestock.

26 Wealth unbound: Andrew Jackson Downing, view of Blithewood estate, New York, 1859. Courtesy of the Foundation for Landscape Studies.

The *Treatise on the Theory and Practice of Landscape Gardening* sold thousands of copies and spawned scores of imitators in the later decades of the nineteenth century. Many writers went even further than Downing in advocating the concealment of lot boundaries. In *Beautifying Country Homes*, for example, Swiss emigré Jacob Weidenmann called fences "decidedly objectionable. When there is no necessity for a fence, do not build one to cut up the land, and define its limits to the spectators. Landowners would have the credit of owning more land than they really possessed could they do away with fences, which always make the property appear smaller than it is."[161] Weidenmann conceded that hedges may be necessary "where unpleasant objects need to be kept out of sight," but "as the house ought not to be cramped in space, and should afford as liberal a view as possible, it is better to remove such things as require a hedge to cover them further away from the house."[162] In a similar vein, Frank J. Scott, whose 1870 *Art of Beautifying Suburban Home Grounds* was dedicated to Downing and went through scores of printings, wrote that "for country, or large suburban grounds, it is safe to say, except where hedges are maintained, that that kind of fence is best which is least seen, and best seen through. . . . Our fences should be, to speak figuratively, transparent."[163] Where fences were unavoidable, Weidenmann advocated the "wire net-work" first produced on English textile looms in the 1840s, whose modern descendant is the chain-link fence.[164] "Though little known at present in this country," he wrote, "they will soon gain the favor they merit. . . . Their durability, lightness, and little cost, place them above all others. Being almost imperceptible, they do not obstruct the view on ornamental grounds, while they possess all the desirable qualities of a good fence."[165]

The argument in favor of transparency was made not only on aesthetic grounds, but also on moral ones. Marking lot boundaries was not just bad taste in the idyll of the early American suburb—it was a "barbarism" that reeked of the rigid class hierarchies and tyranny of the Old World, a mark of that "old stone savagery" Frost would allude to, only several decades later, in "Mending Wall." Scott was the most vocal in expressing this sentiment, instructing his readers that "the practice of hedging one's ground so that the passer-by cannot enjoy its beauty, is one of the barbarisms of old gardening, as absurd and unchristian in our day as the walled courts and barred windows of a Spanish cloister, and as needlessly aggravating as the close veil of Egyptian women."[166]

The attitudes of Downing and his followers were profoundly influential in shaping the landscapes of the first American suburbs. The most notable of these was Llewellyn Park, New Jersey, built by drug merchant Llewellyn Haskell in 1852. Haskell had purchased a dramatic parcel of land above the Passaic River, thirteen miles from Manhattan and located on the Delaware, Lackawanna, and Western Railroad. He hired architect Alexander Jackson Davis, author of the 1837 *Rural Residences* and a close friend of Downing's, to plan the subdivision of the land into residential lots. (Downing was himself originally involved in the design of the plan, but he died before its completion.) Davis laid out unusually large, irregularly shaped parcels along roads that curved to accommodate the topography and natural features of the site. The most distinctive part of Davis's plan, which the architect Calvert Vaux would call "the most sensible real estate development in American history," was the fifty-acre "Ramble," a common area left entirely in its natural state.[167] The Ramble created the illusion of an untouched landscape at the very center of the composition, even as private lots radiated discreetly from its perimeter.

To preserve this illusion, the bylaws of Llewllyn Park provided specific guidance on how parcels could be marked. Though landowners were allowed to design their lots according to personal taste, "every effort was made to harmonize each site with the natural fall and character of the land."[168] These efforts included an explicit ban on the construction of fences, with individual lots left unmarked, the sweeping lawn of one "cottage" flowing into that of its neighbors or defined by shrubbery and trees as Downing had instructed. The result appeared to be not a series of residential properties, but a single common holding. The author of a contemporary review of Llewellyn Park noted with approval that "a number of the holders of lots, entering into the spirit of the place and the design, intend to improve their lots with reference to each other and the whole enclosure, so that the appearance of one large estate may be suggested."[169] And one early resident, Theodore Tilton, described "each estate being isolated from the next, yet each, by a happy partnership with every other, possessing the whole park in common, so that the fortunate purchaser of two or three acres becomes a virtual owner of the whole five hundred."[170] The absence of fences meant that a landowner could lay claim to both his own parcel and the common at the same time.

From the beginning, Haskell conceived Llewellyn Park as a utopian community for the well-to-do. He belonged to a group known as the Perfectionists, who held that "by correct living they might attain the perfect existence on earth."[171] Downing's advocacy of the gardenesque seemed to accord perfectly with this notion, the cooperative "partnership" among residents of the development embodied in the open lawns and sweeping views of its landscape. In the first years of the development, Haskell's vision was not only inscribed in material but reflected in the practices of residents, who celebrated May Day in the Ramble, attended a community "Lyceum," and maintained a common greenhouse.[172]

This "happy partnership" was hardly available to all, however. Entry into Llewellyn Park was restricted to residents and their guests, who had to obtain written permission and pass through a manned gate at the edge of the development.[173] Ironically, this air of exclusivity attracted residents who had little interest in Haskell's utopian vision. If the first settlers of Llewellyn Park had been abolitionists, poets, painters, and editors, many of them personal friends of Haskell, later buyers included the Gilded Age oligarchs Thomas Edison, George Pullman, and Elisha Otis.[174] Over time the Lyceum and greenhouse closed, the May Day celebrations ended, and the only remaining affinity among the residents of Llewellyn Park was the common cause of extreme wealth. The absence of internal boundaries may still have suggested a community living in harmony with nature, but the idyll was one of privilege. The newer residents of Llewellyn Park did not need to mark the boundaries between their parcels because, like holders of neighboring estates, their interests largely coincided. At the same time, the outer perimeter of the develop-

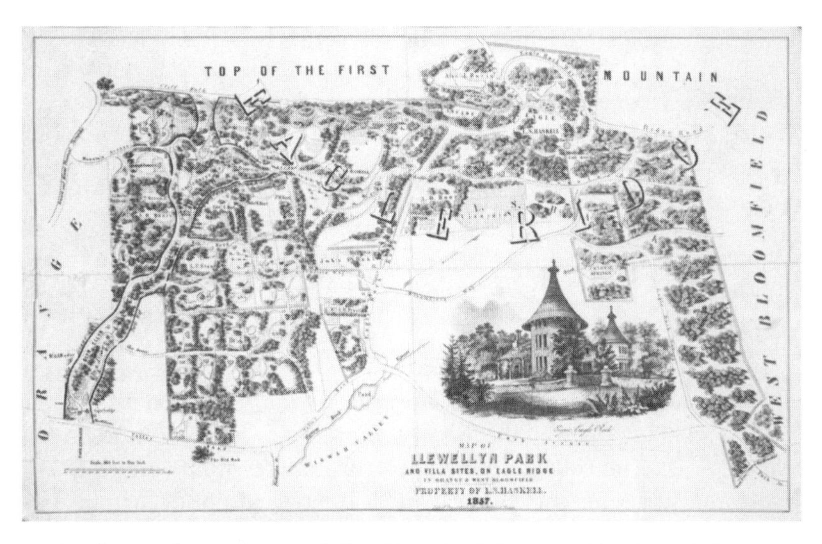

27 Fenceless paradise: gatehouse and plan of Llewellyn Park, Orange, New Jersey, 1857. Courtesy of the Metropolitan Museum of Art.

ment, with its formal gatehouse and dense plantings, was a clear statement of social power and exclusion.

The design of Llewellyn Park influenced scores of new suburbs in the years that followed. Many of these developments forged the same link between wealth and openness that had characterized Davis's design. Tuxedo Park, a combined country club and housing development built north of New York in 1886, was enclosed by a wall eight feet high and twenty-four miles long and manned by a private police force, but fences and walls were banned between parcels in its interior.[175] A decade later, the new railroad suburb of Riverside, Illinois, whose design by Frederick Law Olmsted was to inspire dozens of subsequent developments, prohibited fences along lot frontage.[176] And after World War II, the developers of Levittown, New York, would apply the aesthetic and moral principles of Downing on an unprecedented scale, in a subdivision designed for the exploding middle class. To sustain the vision of what they called "houses in a park," the developers banned fences, walls, and hedges altogether, stating in the by-laws that "fences may not be erected without permission from Levitt & Sons. Even if you get authority to erect a fence, consult your neighbors before you go ahead. Many fences have become complete barriers to friendship."[177] A later edition contained a similar prohibition: "Fences are restricted by covenant. . . . Be a good neighbor and a wise citizen. Do your share in keeping Levittown a 'Garden Community.'"[178] The lack of fences was to produce an overall impression of comfort, ease, and sociability, a vision of the great class of property holders that Jefferson had foreseen two centuries before. The only difference was that now this vision was to be demonstrated not by the presence of walls, but by their enforced absence.[179]

Parcel and Common

In 1937, a century after the enclosure of Otmoor and the publication of Loudon's *Suburban Gardener* and Downing's *Treatise*, the English writer Clough Williams-Ellis published a book of essays with the title *Britain and the Beast*. The collection included contributions from some of England's most prominent authors, intellectuals, and politicians, including economist John Maynard Keynes, authors E. M. Forster and C. E. M. Joad, and historian G. M. Trevelyan. The book was prefaced by, among other luminaries, former prime minister David Lloyd George.

The "beast" in the title was unchecked urbanization in the years after World War I, as "ribbon development" along the main roads out of London, Manchester, and other English cities pushed into the countryside, a process Williams-Ellis said had left England a "mutilated corpse."[180] The typical ribbon development was a speculative enterprise by owners of land abutting the road, who subdivided their property into lots on which they built duplexes with small gardens.[181] As these developments grew, critics argued, they degraded the countryside and severed the link between road and open country. Williams-Ellis illustrated his revulsion with this exchange from Evelyn Waugh's 1930 novel *Vile Bodies*, in which two characters take in the landscape from an airplane:

> Ginger looked out of the aeroplane: "I say, Nina," he shouted, "when you were young did you ever have to learn a thing out of a poetry book about: '*This sceptered isle, this earth of majesty, this something or other Eden?*' D'you know what I mean? '*this happy breed of men, this little world, this precious stone set in the silver sea. . . .*
>
> "'*This blessed plot, this earth, this realm, this England . . .*'"
>
> Nina looked down and saw inclined at an odd angle a horizon of straggling red suburb; arterial roads dotted with little cars; factories, some of them working, others empty and decaying; a disused canal; some distant hills sown with bungalows; wireless masts, and overhead power cables; men and women were indiscernible except as tiny spots;

they were marrying and shopping and making money and having children. The scene lurched and tilted again as the aeroplane struck a current of air.

"I think I'm going to be sick," said Nina.[182]

The essays in *Britain and the Beast* were highly influential in spurring the English regional planning movement and the development of so-called new towns after World War II.[183] One of the most striking things about the book, however, was not the essays between its covers but the cover itself. It showed a stylized open landscape of hills and valleys meant to recall, no doubt, the open heath of Otmoor or Dartmoor. Marching rudely across the image, from foreground to background, was an apparently endless iron fence with menacing palings. The fence receded into the distance and disappeared over the horizon, where there stood what looked to be one of the selfsame "bungalows" that had provoked such nausea in Nina. The viewer stood on the opposite side of the fence from the house, creating the sense that all the land behind the fence had been enclosed by the distant inhabitant.

Here, then, was the modern incarnation of Thomas More's "single fence," the attempt by a single owner to monopolize and privatize a common inheritance. Unlike in the sixteenth century, however, the author of this rapacity was not the hereditary landowner (the editors might just as easily have chosen the image of an estate) but a new class of rural magnates that one contributor, H. J. Massingham, referred to as the "man in the motor car, the new townsman in the old manor house . . . the owner of the red-brick villa with deal boards nailed on to its gables."[184] He continued:

If we look back a hundred years, it will be to view the Industrial Revolution whose fruits are our sour grapes completing the work of the enclosures. The new manufacturing towns were creeping over the northern countryside like a fungus over the leaves of a plant, and the thirst of steam and steel and other newborn giants of industrialism was accelerating the drainage of the countrymen off their land.[185]

The image of the fence suggested the loss, then, not just of open fields or commons, but of the entire way of life that characterized English rural culture before the Industrial Revolution. Physical enclosure was the embodiment of the sovereignty, no longer of kings or earls, but rather of the urban industrial capitalist, the subdivider, the speculator preying on his native land.

28 "Single fence" triumphant: cover of the anthology *Britain and the Beast*, 1937.

The cover of *Britain and the Beast* captured in a single image the tension in the symbolism of walls and fences during the past two centuries. If Locke, Young, and Jefferson had viewed fences as salutary expressions of the improvement that built property and nations, John Clare had seen them as the instruments by which a common resource was removed, often violently, from general use. Loudon and Downing viewed fences both as coarse reminders of the existence of property and as the objects that prevented the illusion of more of it. And both Scott and Frost had seen fences and walls as nothing short of holdovers from the barbarism of the Old World. These ideas have shaped the ways people think about and build boundaries to the present day.

There were ample reasons for this change in symbolism. Much of the idea of "improvement," the principal justification of property that Locke had set forth in the *Second Treatise*, turned out to mean taking land out of the hands of some and putting it in those of others. Often, as was the case with parliamentary enclosure, the "common" Locke wrote about had been governed by use rights that went back many generations, rights that were in point of fact "older than any manorial lord's."[186] Within a few decades, however, the common began to be subjected to the *exclusive* rights of property, with its new owners fencing it and defending it with violence if necessary. A similar transition marked the settlement of the American West, where subdivision on a continental scale was eventually abetted by a fencing material that made it possible to assert a claim on land, irrespective of any right conferred by law, rapidly and without political deliberation. The result was a battle for sovereignty between settlers and the large "interests" that would eventually gain almost complete control over the economic and physical landscape of the United States during the Gilded Age.

It is not hard to understand, in this political and economic context, why a parallel aesthetic discourse about walls emerged in the United States. This discourse, epitomized by Downing, sprang from a national romantic desire to distinguish the new nation from its corrupt European parents. Physical boundaries were deemed unseemly and "barbaric" expressions of political and economic inequality, and as such should be avoided or, at the very least, concealed. Early suburban developments like Llewellyn Park were predicated on this idea, with the lack of fences suggesting a utopian community residing in harmony with nature. It is arguably from this aesthetic aversion to fences that there emerged, slowly and fitfully, the wider idea that walls were inherently antithetical to democracy and community—a view that persists in the United States today. Walls and fences were reminders not just of settlement and improvement, but of the appropriation of a vast public domain whose existence made America superior to its European progenitors. They were inextricably implicated in the loss of what was distinctly *American*. As they did for the contributors to *Britain and the Beast*, walls and fences marked a fall from a golden age of openness.

However tempting this view may be, it is important to recall the roots of the aesthetics that Downing and his followers advocated. History shows that the impulse to conceal boundaries originally had little to do with democracy or community, but rather lay in landowners' desire to naturalize the contingent set of economic relations that developed in England in the late eighteenth century, in which they suddenly enjoyed unprecedented advantage. No less than the new fences and hedges in the outer fields, the invisible ha-ha around the formal garden sent a message about "where the power now lay."[187] Downing's aesthetics were never far removed from these origins. A moralist and an aristocrat at heart, Downing was finally no less concerned than Brown and Repton had been with maintaining the appearance of wealth and ease. Just as at Blenheim, the lack of visible boundaries in Llewellyn Park served to naturalize a particular system of property relations in which the country's new industrialists lived in "country places" at the end of commuter railway lines while the poor and the immigrant inhabited "narrow cells" in the city.

These are the roots of the fundamental tension in the way people think about walls today. Walls continue to be powerful symbols of civility and community—the white picket fence retains unambiguously positive associations in American culture—but they are also seen as the instruments by which the owners of property, whether a large landlord or a suburban home owner, express and assert the set of rights that derive from that ownership. This tension between two visions of the wall, that of Locke and that of More, can be reduced to an essential dichotomy. On the one hand, the *parcel* with its rights of property; on the other, the *common* with its obligations of use. It is usual to conceive the poles of this dichotomy in absolute terms. Either a wall turns the common into a parcel (Locke), or the lack of a wall turns parcels into a common (Downing). In this formulation, a gain for the parcel is a loss for the common and vice versa; there is scant middle ground between the two. Physical boundaries are today nearly ubiquitous and highly diverse objects, but this diversity is barely reflected in the debates about them. Just as in the late seventeenth century, a boundary is either an enforcer of property or the place where every trace of property is obliterated.

Yet there is another way. It is possible to see the walls and fences of the modern landscape not simply as expressions or enforcers of property, but as unique places where the prerogatives of parcel and common interact. Physical boundaries are sites of active and ongoing social production, places where the rights of the group and the rights of the individual are negotiated, whether through deliberation or through conflict. In other words, walls are places where important work gets done. Adhering to the notion that walls are little more than "barbaric" emblems of power precludes thinking about them in this broader, socially productive way. It is also not constructive in a subdivided and diverse landscape where walls are going to be built, whatever one thinks of them. In the modern landscape it cannot come down to a choice of *either* parcel *or* common. It must always be *both/and*.

More than any single object, the wall itself—that thing of wire, wood, stone, or plants—is where this resolution must take place. The road to recovering the wall starts with ceasing to see walls as constructions of sovereignty and returning to what they once were.

FOUR

Recovering the Wall

On a cold, drizzly night in late November 1989, strange images began to drift in over the airwaves from Berlin. The new East German government had decided some days before to allow travel to the West with a visa, but in the late afternoon of November 9 it had suddenly announced that henceforth travel outside the country would be unrestricted.[1] Television footage from only several hours later showed incredulous East German citizens streaming across Checkpoint Charlie, the famous border crossing between the American and Soviet sectors, while East German, American, and Russian officials looked on. Then even more striking images started to pour in. They showed crowds massing along both sides of the Berlin Wall in front of the Brandenburg Gate, some using hands and shoulders to climb onto the wall itself. Several husky men, armed with sledgehammers that appear as if out of nowhere, bang away at the thick concrete slabs. A chink appears in the wall and grows bigger and bigger with each blow. Finally a young man manages to squeeze through the hole, and a delirious roar erupts from the crowd on the western side. Now hundreds of people are standing on top of the wall, shouting and laughing and drenching the crowd below with champagne. The work with the sledgehammers continues as more and more people pass through the wall. And with every person who pushes through, every concrete chunk that falls, it is as if one more piece of a world order is shattered.

The images recorded that night are among the icons of

the twentieth century. Looking back at them today, one can almost feel the weight of history being lifted and the sense of possibility rushing in to fill the void. The fall of the Berlin Wall seemed to usher in a new political era before the very eyes of those who witnessed it; never before in modern times had an apparatus of state control collapsed so suddenly and dramatically, and with such immediate physical effects. "I remember that night as a kind of dream," recalled one who was present. "The atmosphere was one of giddy excitement and joy but also one of sheer disbelief."[2]

And yet, historically speaking, the Berlin Wall had a very short life. In the early 1960s, the government of Erich Honecker was growing increasingly alarmed by the flow of East Germans across what was then an open border with the western half of Berlin. At first residents had traveled to see family and buy goods that the West German economy, exploding after the devastation of World War II, was producing. Over the years, however, more and more of these "visitors"—whose numbers included many scientists, technicians, and teachers—did not return. In 1960, nearly 200,000 East German citizens left the country.[3]

The Politburo decided to halt this exodus, which threatened to eviscerate the East German economy once and for all. On the night of August 12–13, 1961, construction workers guarded by East German police began stringing barbed wire and building a rudimentary cinder block barrier along the boundary between Soviet-controlled East Berlin and the American, English, and French sectors in the west. Later, in the center of the city, this barrier was replaced by fifteen-foot-high concrete panels with horizontal cylinders on top to prevent climbing over. The wall was paralleled on the East German side by a floodlighted trace road and killing ground with manned watchtowers every few hundred feet. The concrete wall through the center of Berlin was only one part of a system of walls and electrified fences that extended more than seventy-five miles, enclosing the entire western half of the city. Thousands of people were captured by the East German authorities trying to cross these fortifications, and it is estimated that at least 136 died in the attempt.[4]

The Berlin Wall remains a universally recognized symbol of political oppression, of the heavy-handed and ultimately futile effort of a state to cow and imprison its population. The wall also embodied the Cold War as a whole, a forty-year contest for dominance on the part of two states capable of destroying each other and the world. It was an almost perfect example of a boundary built to express sovereignty and block contact, and its fall was rightly celebrated around the world.

But its fall is not the only story that can be told about the Berlin Wall. During West Berlin's isolation, the West German government provided subsidies to residents of the city as a way of sustaining economic life there. These subsidies were of questionable economic value, but they made West Berlin a haven for artists, students, radicals, and immigrants. The wall was a magnet for many of these diverse communities. In a city where public plazas still lay in ruins, the open signboard and outdoor museum of the Berlin Wall quickly became one of the most actively used spaces of the city. For nearly three decades, much of the cultural and economic life of West Berlin unfolded in the real or metaphorical shadow of the wall. Because one could approach the wall on the western side without restriction, there was always some event, some gathering, some demonstration going on somewhere along its length.[5] Crosses were erected to commemorate those who had died trying to breach the wall, turning it into a permanent memorial even before its destruction.[6] And some have even argued that the wall *increased* rather than decreased political stability.[7]

The most famous use of the wall was as a signboard for all manner of messages; by 1989 its concrete panels had been completely covered by graffiti. The messages ranged across the political spectrum, from attacks on East German and Soviet political oppression, to mockery of West German materialism, to denunciation of American imperialism; perhaps the most famous image showed Leonid Brezhnev and Erich Honecker locked in a lovers' embrace. In the later years of its existence, the wall itself became a major tourist attraction, and thus responsible for a significant part of whatever economic vitality Berlin had.[8] In a divided city ravaged by war, one whose former identity as the capital of the Third Reich was a source of embarrassment and shame, the Berlin Wall was, for a brief moment in history, the *genius loci*—the very spirit of the place.

For this reason, what to do with the wall was a matter of considerable debate in the months following November 1989. Should all traces of it be erased and Berlin brought back to its prewar state? Should sections of the wall be left in place as a monument? Or should some other means of remembering the wall be adopted? A small but consistent minority argued that the wall was a vital part of Berlin's political, economic, and cultural history and should be retained in some form. For example, the journalist Jochim Stoltenberg argued in the *Berliner Morgenpost* that "a few meters of Wall should remain standing as a memorial. That may be painful to some, but this decision is unavoidable. This structure of con-

29 Memento of a moment: Berlin Wall, Berlin, Germany, 2012. Courtesy of Ida Pedersen.

crete and barbed wire has caused too much inhumanity and too much suffering . . . for its complete removal to be warranted."[9]

As it happened, most of the Berlin Wall was quickly demolished, its remnants trucked off to join the rest of Berlin's war rubble on the edge of the city, smaller pieces sold off to tourists from around the world. Today the course of the Berlin Wall is marked with benchmarks in the pavement, and the one section of the wall still standing—the East Side Gallery, along the Spree River in the neighborhood of Friedrichshain—remains one of the city's chief tourist attractions. Today the wall is not just a monument of the Cold War and a symbol of tyranny; it is also the most important single vestige of the culture of artistic and political experimentation of Berlin in the 1960s and 1970s—a culture many fear is being lost with the city's conversion from radical outpost to national capital.

The unique political and material history of the Berlin Wall shows how, even when walls are built to forcibly separate people, it is hard to keep them from assuming other social functions, some of which may even lead to the destruction of the wall itself. This is not only a matter of history. There are many Berlin Walls, and one need only look carefully to begin to see them.

Resisting Sovereignty

On a warm evening in April 2004, two groups of protestors gathered on either side of the high concrete panels that bisect the Abu Dis neighborhood of Jerusalem, one small part of the elaborate system of fences and checkpoints that Israel built in the West Bank during the Second Intifada. They had come for a novel demonstration of what one can do with a wall, a wire, and a couple of video cameras. The protestors set up cameras on either side of the wall and wired each to a projector on the opposite side, taking advantage of the regular weep holes punctured at the base of the wall to allow for the flow of water. When the video cameras and projectors were switched on, they produced what one witness described as "a very large virtual hole in the wall." The two groups had suddenly become one, "singing, dancing, and cheering as though the wall was not there."[10] Reporters from both sides were on hand to capture the scene.[11]

This event was organized by Artists Without Walls, a group of Israeli and Palestinian artists that "seek[s] to eradicate the lines of separation and the rhetoric of alienation and racism . . . through nonviolent and creative actions."[12] The group was cofounded by Israeli photographer Miki Kratsman. Kratsman had been documenting conditions along the wall during the entire period of its construction in the early 2000s. He was particularly interested in documenting small gestures—graffiti messages, a giant reproduction of Michelangelo's *Creation of Adam*—that artists left on the new barrier to contest the political and social assumptions the wall embodied. Early in this work, Kratsman had noted that the wall's smooth, blank surface functioned perfectly as a screen for showing videos and movies.

After making the "virtual hole," the members of Artists Without Walls began to organize yearly protests at the same site, all using the same method of momentarily erasing the wall to unite the groups on either side. At one event, protestors played tennis over the barrier; at another, they engaged in a communal drum-beating exercise where, as described by the artist Adi Louria-Hayon, "Darbukas, Jericans, pots, and lids all worked together, to combine sound in a noble effort to dismantle visual and auditory barriers."[13] If only temporarily, the events organized by Artists Without Walls transform a concrete barrier into a stage for social

30 The wall undermines itself: "virtual hole" demonstration, Abu Dis, Jerusalem, 2004. Courtesy of Dean MacCannell.

exchange and political agitation. As one participant in the 2004 protest wrote, "With a prodigious act of the imagination, even this most forbidding wall can be used as a device to bring people together."[14]

Like the demonstrators who routinely gathered along the Berlin Wall, residents of Jerusalem have thus seized control of the messages the "seam line obstacle" sends to the world. From a symbol of separation, the wall has become a symbol of the fundamental common interests of Jew, Muslim, and Christian. This hints at a significant irony. Although the wall has made life extremely difficult for Palestinians unlucky enough to live in its path, it has also focused world attention on the rights of Palestinians in a way that was never possible before. Some have noted that a striking physical barrier running through an urban neighborhood, an image that inevitably recalls Berlin, has even helped their political cause. The wall provided, suddenly, a focus of opposition, some *thing* to protest, a rallying and gathering point, and a surface on which to project the very messages that undermine it. Accordingly, some fear that were the wall to come down as suddenly as in Berlin, it would make broadcasting these messages to the world far more difficult.[15]

Attempts to seize control of the messages a wall of sovereignty sends are also being made by artists and citizen groups at the fence the United States has built along one-third of its border with Mexico. In Tijuana, the Mexican city immediately south of San Diego, the boundary, which along this stretch is a wall of corrugated metal, has been turned into a virtual outdoor museum and gallery space, a canvas that draws artists from all over Mexico. In one section, the multimedia artist Silvia Gruner mounted "111 identical plaster figurines representing the Aztec goddess of filth and purification, Tlazoltéotl, grimac[ing] in the pain of childbirth."[16] At another, Alberto Caro turned the fence into a memorial for those Mexican migrants who have died in the desert to its east, whose numbers have risen in the years since its construction. Mounted to the fence are a series of vividly painted coffins, each bearing a large black number. Above each coffin, a small panel shows the year. The installation thus provides a running tally of deaths for the thousands of people who pass it every day. Another installation—by the artists Susan Yamagata and Michael Schnorr in collaboration with the Coalición pro Defensa del Migrante, a Tijuana-based migrant aid organization—is composed of 5,100 white crosses affixed to the fence along the road to the city airport. Along this stretch Yamagata made a giant papier-mâché boot with a skull motif bearing the question *¿Cuantos mas?* "How many more?" She also worked with San Diego artist Todd Stands to paint three virtual doors in the surface of the fence; the one "open" door shows a view of the desert beyond. Farther east, in Nogales, a city divided between Arizona and the Mexican state of Sonora, a collective of artists and activists has mounted metal sculptures and painted murals on the fence. These gestures are paralleled by the "quieter musings" of migrants whose names and places of origin have been scratched into the metal.[17] In all these cases, the fence has been used as a signboard for challenging the messages of sovereignty and power it was designed to send. It is a way of turning a wall of control into a wall of belief, ritual, memory, and politics.

The artists and activists working along the United States border fence operate under constraints far greater than those of Kratsman and his group. Unlike the situation in Jerusalem, where access to the wall is unrestricted on either side, the border fence is impossible to approach on the United States side, since the entire zone to its north is overseen by the Border Patrol. The only places where it is possible to alter the fence physically are therefore those urban neighborhoods where it sits exactly on the political border between the United States and Mexico. As in Berlin, where the eastern side of the wall was a killing ground, appropriating the structure of a wall to change the messages it sends can happen only where it is possible to reach the wall itself. This is why the builders of walls often want to keep the public well clear. For example, in the

31 Boundary of remembrance: "Muertes Mexicanos," Tijuana, Mexico, 2007. Courtesy of Luis J. Jimenez/New York Times/Redux Images.

ironically named Friendship Park south of San Diego, it is no longer possible to approach the border fence that separates the United States from Mexico, since an even higher fence has been placed hundreds of yards to the north to create a "security zone" between them. As elsewhere in San Diego and Tijuana, the wall here is now a social space only on the Mexican side.[18]

It is not only new national walls that are actively contested all over the world, but also walls reaching back many centuries. This is the case in Britain, where grassroots associations work to reopen, and keep open, rural paths blocked by private walls and fences, some dating from the era of John Clare. Since the early part of the twentieth century, these organizations have turned "trespassing [into] a mass movement" in England.[19]

One of these associations is the Ramblers, a group devoted to protecting the rights of walkers in the landscape. The group works to maintain rights-of-way across all manner of boundaries throughout England and to give walkers advice about their legal right to traverse open access land.[20] These practices are the modern continuation, more than a century later, of the fundamental conflict that emerged with the shift of the

English countryside from use rights to property rights in the eighteenth and nineteenth centuries. In many places common law still protects the public's right to use paths, provided they are walked at least once a year. If a year passes without at least one documented walker, a landowner may close the gate or stile in a fence or wall, preventing access not only to her property but to the entire path. It is thus important that all public rights-of-way be walked regularly, and that this walking be recorded in a public register.[21] One of the prime activities of the Ramblers is therefore to coordinate these walks all over England. The group organizes yearly events for increasing general awareness of England's extensive network of public rights-of-way and encourages people to exercise their common law rights to walk paths, open gates, and climb over walls. The Open Spaces Society, the other major group devoted to defending and promoting rights of access in the countryside, retains a list of all the paths subject to common rights of access. Both groups make people aware that the sovereignty of property is not absolute, and that they retain rights to the landscape that they may never have known they had. These groups were directly responsible for the drafting of the Countryside and Rights of Way Act of 2000, which finally enshrines in national law the rights of access they have fought for over the years.[22]

Maintaining access to rights-of-way can have profound effects on the physical form of boundaries. This is perhaps best seen not in the countryside, but in urban neighborhoods throughout England, where a movement has recently emerged to contest "gating orders." In many neighborhoods, local authorities have closed walking paths to deter criminal activity. However, this is often done without consulting residents about the effect such closure will have on traditional practices and patterns of use. Since 2006, when the legal mechanism for gating orders was introduced, approximately 1,500 public paths all over the United Kingdom have been gated or otherwise blocked; to date there has been no case where a gate that was installed was later removed. The Ramblers have thus begun a campaign to "have the law amended so that when a member of the public does object to a gating order the decision is passed to an independent decision-maker."[23] They bring attention to gating orders that have significant negative effects on public life by limiting access to schools, parks, and shops, or where footpaths provide a safer walking route than roads; contest the closing of gates without proper public debate; and report violations under the Countryside and Rights of Way Act. In short, the Ramblers do not object to the *existence* of walls or fences, but rather protest the degree and location of their permeability in response to human practices and needs.

32 These paths are unstopt: "Ramblers" asserting rights of common access, West Sussex, England, 2011. Courtesy of Sue Sinton Smith.

Such attempts to secure public access are all ultimately based on the notion of use right. A use right gives people access to resources without their actually "owning" them in the modern legal sense. Such rights were increasingly circumscribed in the eighteenth century, but now they are being vigorously defended in England and elsewhere. However, not everyone has access to this legal instrument when contesting a wall. The work of the protestors at Abu Dis and Tijuana falls instead under the heading of what French philosopher Michel de Certeau called "tactics." According to Certeau, tactics "use, manipulate, and divert" spaces and "create a certain play in the machine through a stratification of different and interfering kinds of functioning."[24] Tactics in this sense are different from "strategies," which "seek to create places in conformity with abstract models."[25] Many attempts to commandeer the messages a wall sends do not fundamentally alter the way the wall operates, but rather take advantage of the inability of a sovereign interest to fully manage its own methods of control.

But even if tactics are not identical to strategy, they can be its first steps. Unlike the situation in Berlin, the appropriations of the Israeli "seam line obstacle" or the United States border fence have been orchestrated as part of wider coordinated political movements.[26] By challenging the idea of national territory as distinct and inviolable, these movements deliberately throw open to debate the very question of what a boundary is and what it may justifiably do.

Rethinking Territory

The hedge is the original wall of the landscape. Its form is the result of interactions over many years between the birth, growth, and death of plants, animals, and microorganisms and ongoing management by people. Once a hedge is established, it is likely to endure longer than almost any other kind of boundary, as the ancient embankments of Cornwall demonstrate.[27] This endurance is one reason hedges have been planted for millennia by human communities across the world, from Japan to England to the United States.

The earliest function of hedges was to control livestock, but today they play an important role in improving the ecological performance of landscapes.[28] Many studies have confirmed that hedges form "habitats, refuges, corridors or barriers [that] are critical for many plants and animals that otherwise could not exist in agricultural landscapes."[29] Hedges "provide a wide range of food for birds of widely differing feeding habits—grass seed for linnets, caterpillars for cuckoos, hawthorn berries for fieldfares, thistles for goldfinches, ash keys for bullfinches, snails for thrushes, earthworms for blackbirds, mice for owls, and small birds for sparrow hawks."[30] Because they are exposed to the sun on both sides, hedges probably house more woodland edge insects than woods do.[31] This habitat function has also provided discernible benefits to agriculture, since many of the vertebrate species that live in hedges prey on pests.[32] Hedges also improve soil conditions by reducing wind speed over large areas; early research from western Denmark showed that prevailing winds blow on average six kilometers an hour slower in areas planted with hedge networks than in areas that lack them.[33]

Yet hedges are under considerable stress in the modern landscape, as the case of England demonstrates. During the era of parliamentary enclosure, the structural diversity of hedges was gradually reduced until they were little more than monocultures of hawthorn, with an attendant loss of many of their ecological benefits.[34] With the advent of modern agricultural methods after World War II, many ancient hedges were eradicated to consolidate fields and make room for larger machinery.

Newer hedges tended to be left in place, both because parliamentary enclosures were larger and because the hedges themselves, structurally less complex than their predecessors, took up far less area. The result of this process was a 50 percent decline in the aggregate length of hedges in England between 1950 and 1990; the reduction was 20 percent between 1984 and 1990 alone.[35]

Beginning in the 1970s, this drastic reduction in the number and quality of hedges sparked a popular movement to restore hedges in England. Defenders of hedges were concerned about the loss not only of their ecological or economic benefits, but also of their cultural associations. The ancient hedges that preceded the rapid landscape transformations of parliamentary enclosure and the industrialization of the English countryside in the late nineteenth and early twentieth centuries have assumed increasing importance as emblems of English identity and heritage. As one researcher writes, hedges are "felt to represent the Englishness of the landscape."[36] Today, practically every cultural heritage advocacy group in England has an official position in support of hedges. Many of the rituals related to hedge maintenance are now the subject of events and competitions across England, including the National Hedge Laying Championships at Fakenham in Norfolk. These popular movements spurred the drafting and passage of the Hedgerow Regulations of 1997, the first piece of legislation in the world directed specifically at protecting, managing, and expanding hedgerows.[37] The result is an apparent halt in the loss of hedges during the past two decades.[38]

Advocates of hedges argue that it is not right to think about them simply as objects like a wall or fence. Hedges are large-scale ecological networks that extend far beyond the boundaries of any one enclosure. Because "no single hedgerow can harbor all the local species pool of a given group [of] plants, birds, or insects," interconnection of hedges is critical for maximum ecological benefit.[39] This is particularly important in areas where forest is scarce, since hedges create corridors for the movement of larger vertebrates. When aggregated, hedge networks can be very large indeed. Even after decades of eradication, in the 1970s the area covered by hedges in Britain still amounted to 400,000 acres, or twice the area covered by national reserves; this corresponded to roughly one-quarter of all deciduous woodland in Britain.[40] These numbers suggest that hedges, even now, remain a critical repository for plants and corridors for animals in the landscape.[41] Hedges can thus be seen not only as boundaries between different areas, but as vital ecologies in themselves. This notion has significant implications for boundaries at much larger scales—even an entire continent.

Over the past several decades, the Sahara Desert has expanded at an alarming rate. The threat of desertification is particularly great in the countries of the Sahel, the enormous dryland zone south of the Sahara that stretches the entire 5,000-mile width of Africa. The southward advance of the desert is threatening almost every aspect of human life in these countries, from agricultural productivity, to political stability, to the very existence of towns and cities. There is thus a pressing need to work across political boundaries to slow or reverse it.

One of the most ambitious attempts to meet this challenge is a new kind of boundary that, if constructed, will span the entire continent. As envisioned by its advocates, the "Great Green Wall" would consist of a nine-mile-wide band containing a diverse array of native trees, hedges, and shrubs that all thrive in the extreme aridity of the Sahel. When complete, this boundary would extend from Senegal to Ethiopia and cover nearly 30 million acres of forests, fields, farms, and villages.[42] The "wall" will therefore be more than a boundary between the Sahara and the farmland to its south: it will be an entirely new ecology, the longest, thickest hedge on the planet.

The work of one organization in Burkina Faso—a landlocked country in East Africa whose capital, Ougadougou, is threatened with desertification—provides a preview of this unique ecology. Nouvel Arbre, "New Tree," worked to create strategic hedge enclosures where native plants could regenerate without being destroyed by overgrazing. Ougadougou is now ringed with a network of such enclosures, which increase soil fertility by reducing wind speed, retaining moisture, and adding organic material. Working with local populations, the group has now begun to plant similar hedge networks on a much wider scale, with encouraging results. The director of the group reports that "after four years of operations, we have managed to sustainably protect more than 150 hectares spread over family lands of two to three, even six hectares. The first sites have become small forests rich in local animal and plant species that communities use to feed and care for themselves."[43] If the Great Green Wall is to be built, it will likely be done in this way, piece by piece, hedge by hedge, enclosure by enclosure.

A similar idea has been advanced by the Sahara Forest Project. This group, which is underwritten by the governments of Norway and Qatar, advocates the construction of a continent-spanning array of "Sahara forest greenhouses" that will form an enormous "hedge" running through the desert. The greenhouses would be a giant living machine consisting of enormous solar reflectors, food and energy plants, and desalinized seawater for irrigation. At the same time that it slows the advance of the desert, this boundary would produce energy, crops, freshwater, and salt. In the words of one proponent, the greenhouses are a "pioneering project that reflects the kind of holistic vision . . . that we need to explore so that we can address the interconnected challenges of food, water, and energy security."[44]

The Great Green Wall and the Sahara Forest Project represent a new kind of wall, one that responds to a natural process cutting *across* national boundaries. To pursue projects of this scale and magnitude therefore will demand unprecedented coordination among people and nations; no hedge, no matter how wide and long, can be sustained without ongoing maintenance and care. If successful, not only will these projects slow the advance of the desert; they will also provide a model for cooperation among diverse communities separated by geography, culture, and language. The construction of the wall will require that nations relinquish part of their own sovereignty to achieve a larger common goal. That is an enormous challenge in the contemporary era, and it remains to be seen whether the nations of the Sahel will rise to it. But someday people may look back and wonder that physical boundaries ever did anything other than stop the march of deserts.

The Great Green Wall and the grassroots efforts to rebuild hedge networks in Britain represent two ways of reconceiving territory. Both are based on the idea that boundaries, whether between fields or across nations, need not necessarily disrupt social and natural ecologies but can be used to support them. They therefore have implications for a wide variety of walls in the landscape—even ones from the remote past.

On a cold night in March 2010, what was once the most important territorial boundary in Europe was suddenly transformed in a brief but spectacular moment of public art. Hadrian's Wall, the bulwark built in the second century CE that runs from Newcastle in the east of England to Solway Firth in the west, had long been of interest to scholars, historians, and tourists as the most important Roman ruin in Britain, but its relevance to the daily lives of those who lived around it was slight. A wall that was in its time not just a barrier, but an entire economy with granaries, storehouses, and hundreds of permanent residents, was little more than an abstraction for most people. The organizers of the event that March evening, Hadrian's Wall Heritage, decided to stage an event that would bring the wall back to life, even when the original reason for its existence had long since vanished, like the cows in Frost's poem.

This event was a light show along the eighty-four-mile length of the wall, commemorating the sixteen hundredth anniversary of the end of Roman rule in Britain. A sequence of five hundred "illuminations," gas beacons spaced at 250-meter intervals, would roll west from Segedunum Fort in Wallsend, east of Newcastle, reach the city of Carlisle forty-five minutes later, and end on the last, fragmentary stretch of wall at the tiny village of Bowness on Solway Firth. At the climax of the event, Hadrian's Wall would be completely illuminated.[45] For the first time since the second century, the wall would be a living place, populated by thousands of people along its entire length. Parades and fireworks were planned at Wallsend and Carlisle to celebrate the wall and its significance for the north of England. Residents and artists from the region were invited to submit ideas for individual illuminations; the organizers would be responsible only for coordinating and timing the vast supplies of natural gas that would be needed to fuel them. Considerable effort went into coordinating the event with the landowners and farmers whose land the wall crosses.

On the night of the event, thousands of volunteers gathered along the wall, from the lowlands near Newcastle to the windy moors above Hexham. At exactly 5:45, after a celebration and rooftop performance, the first "illumination" was ignited. Then, a minute later and 250 meters west of that, visible to the crowd below, the group assembled at the next point lit their beacon. Guided from a helicopter flying along the length of the wall, this process continued all the way across England for nearly an hour, the wall gradually lighting up against the night sky. At every lighting, the visitors and residents who had gathered at that point cheered and lit candles, their faces rosy in the light of the flames. One volunteer said: "Hadrian's Wall is unique. This has never, ever happened before, might never, ever happen again." Another described the effect as nothing short of "magical."[46]

In the 1920s the German playwright Bertolt Brecht coined the term *Verfremdungseffekt*, "defamiliarization," to describe his plays, which were often characterized by direct address to the audience, harsh lighting, songs, and explanatory placards. These devices, Brecht said, were designed to "strip . . . the event of its self-evident, familiar, obvious quality and creat[e] a sense of astonishment and curiosity about them."[47] Defamiliarization distances people from the automatism of the known

33 Old wall, new light: "Illuminating Hadrian's Wall," Northumberland, England, 2010. Courtesy of Hadrian's Wall Heritage.

by making them see their world as "strange" (*fremd*). While Brecht used the notion to explain his own particular art, defamiliarization is arguably the foundation of all artistic creation.[48] Art makes what people had ceased to see visible again by distorting it, changing it, or representing it in a new way. This is what happened that night in the north of England. Illuminating Hadrian's Wall turned an archaeological object into social space, a stage for exchange and performance on a territorial scale. It brought to those who witnessed it a fundamentally new appreciation of a familiar place.[49] By showing it literally in a new light, the illuminators of Hadrian's Wall brought it to life again after sixteen hundred years.

The Hadrian's Wall illumination made strange an existing wall, and in doing so it gave people a radically different view of their landscape. But it is also possible to defamiliarize landscape *itself* with a wall. This is what the artists Christo and Jeanne-Claude did in 1976 north of San Francisco, when they constructed *Running Fence* across nearly the entire width of Sonoma and Marin Counties, from Petaluma to the Pacific Ocean.

Since the creation of *Valley Curtain* in Colorado in the early 1970s, Christo and Jeanne-Claude had been interested in constructing new kinds of landscape boundaries. Christo's drawings of the *Running Fence* project as he envisioned it showed a sinuous line of billowing white nylon sails, eighteen feet high by sixty-five feet wide and mounted on steel poles, marching up and down the grassy slopes of the Marin mountains and finally disappearing under the waves of the Pacific. In its entirety the fence would extend twenty-four miles, over private land and public, hill and dale, forest and field. It would not follow any recognized legal boundary at all, but rather would trace the topography of the area.

This flouting of legal and property boundaries made construction far more complicated than the pair of artists had imagined. Erecting the fence required an enormous amount of negotiation with the fifty-seven landowners over whose property it would run. During the years before the fence's construction, Christo and Jeanne-Claude attended seventeen public hearings, the tone of many hostile and uncomprehending. A "Committee to Stop the Running Fence" was formed.[50] At one point during construction, the fence was vandalized.[51] In all, laying the legal and political groundwork for an art installation that was to last only two weeks took nearly five years.[52] Even at the very last minute, as construction of the fence was beginning, its legality where it crossed the land of the California Coastal Commission remained in question.

34 Strange transgressor: *Running Fence*, Sonoma County, California, 1976. Courtesy of Wolfgang Volz/laif/Redux Images.

When it was finally completed in September 1976, the *Running Fence* attracted thousands of visitors from near and far to witness it firsthand. Werner Spies, writing at the time, described the effect of the fence as a "dreamy arabesque, as volatile as quicksilver in the midday heat."[53] One art critic called it "an unimaginably white calligraphy over the hills."[54] Some stopped their cars where the fence ended at Highway 101, admiring the gigantic construction that seemed to have landed from nowhere. Others gathered on the beach where the fence made its final plunge into the Pacific. Still others walked the entire length of the fence, ignoring the fact that the private land it stood on was explicitly closed to public access.[55] The wall quite literally gathered up the landscape around it, as soil and burrs blown by the wind were caught up in its fabric.[56] One observer later called it "one of my strongest encounters with art."[57]

But the *Running Fence* was more than an object to be admired in its own right; it also spawned a variety of events and celebrations along its length during the mere two weeks of its existence. It was a profoundly *social* space, with planes and helicopters buzzing overhead while a small army of "hippies and rednecks" worked frantically on its construction.[58] After its deconstruction, the pieces of the fence were distributed to the people whose land it had crossed.[59] At the final hearing to determine whether the fence would be built, Christo offered an eloquent summary that placed the fence squarely within this social ecology: "The work is not only the fabric, the steel poles and the Fence. The art project is right now here. Everybody here is part of my work if they want it or don't want it. I believe very strongly that twentieth-century art is not a single individualistic experience. It is a very deep political, social, economic experience I live right now with everybody here."[60]

The *Running Fence* transformed the way people looked at received boundaries around them. By creating an entirely new boundary, one that responded to its own internal logic rather than to existing political or property lines, Christo and Jeanne-Claude defamiliarized a physical and social landscape people thought they knew. The fence created a new and entirely different narrative of landscape, one that emphasized not separation, but connection. Rather than dividing, it united. By linking different climatic zones, soil types, property owners, and land uses, the fence created a whole new way of apprehending and experiencing the place where it rose. As one critic has recently written, speaking in the present as though the work still stands, "The *Running Fence* divides nothing, confines nothing, impedes nothing, prevents nothing. It liberates the very concept of a fence, makes it blissfully useless, and frees it from its analog historical negatives."[61] It did this in a way that redefined and revealed the landscape around it. But it is not only at large scales that such defamiliarization can take place. The lesson of the *Running Fence* is equally apposite in the modern urban landscape.

Redefining the Block

In his 1964 book *Investigations in Collective Form*, the Japanese architect Fumihiko Maki noted that "there is a complete absence of any coherent theory [of architecture] beyond the one of single buildings. We have so long accustomed ourselves to conceiving of buildings as separate entities that, today, we suffer from an inadequacy of spatial languages to make meaningful environment."[62] Maki's response was to develop an architecture of "collective form" that would transcend this preoccupation with the single building. Maki was particularly attentive to the relation between different scales of organization in the city, which he viewed as both a whole physical structure and a bewilderingly complex array of material and social details. "At certain moments in our urban lives," he wrote, "we relish all the diversity and disjointedness of cities, and bask in the variety of them. . . . But when a plethora of stimuli begins to divert us from receptive consciousness, the city renders us insensible. . . . If urban design is to fulfill its role, it must do more than simply organize mechanical forces, and make physical unity from diversity. It must recognize the meaning of the order it seeks to manufacture, a humanly significant, spatial order."[63]

At the center of this "order" was the wall, which Maki placed first among components of "collective form." Maki defined a wall as "any element which separates and modulates space horizontally. Walls are places where forces outward and inward inter-act, and the manner of the inter-actions define[s] the form and functions of the wall."[64] This conception of the wall as a separate element that stages interactions between centrifugal and centripetal forces can be seen in many of Maki's designs. The outside wall of the Osaka Aquarium, for example, is a glazed double portico supported by tapering columns that seem to float away from the ground. This structural complexity marks the wall as distinct from street outside and building inside, rendering it a separate, freestanding object in the urban landscape. Maki's interest in walls that interweave "inside" and "outside" is also evinced in his design of the Hillside Terrace housing complex in central Tokyo. Here Maki used raised platforms that extended into the street, overhanging stairwells, and storefronts penetrating the block to blur traditional boundaries between street, building facade, and block interior.[65]

But it is a purely speculative wall that best illustrates Maki's ideas about the way boundaries should function in the urban landscape. In the final pages of the *Investigations*, almost as an afterthought, Maki turned his attention to what he called a "community wall," or a "device to provide a transitional space between busy streets and quiet residential sections. This community wall, which is a continuous mound, is made up by a series of garages, small community stores, gates, children's play area, etc. This is an environmental wall which begins to be molded by activities outside and inside the community."[66] This description was accompanied by a diagram showing a city block whose perimeter was partly lined by small, permeable structures that implied different functions and users. Each shop, garage, and apartment went directly through the wall, and one might gain access to the center of the block through any one. Despite the schematic nature of the drawing, Maki's interest clearly lay in exploring the perimeter of the block as a critical, yet often overlooked, place of social and commercial interaction.

Maki's idea was clearly rooted in the urban patterns of his homeland. Traditional Japanese cities, like their Chinese forerunners, were composed of "wards," or walled urban neighborhoods whose gates were closed at night. But the official gate was not the only way to enter a ward. Maki's "environmental wall" bore a strong resemblance to the *machiya*, the traditional storefronts that enclosed residential precincts in Japanese cities before the nineteenth century. The *machiya* was a low, deep, and narrow structure where activities of living and selling went on under the same roof. From the street, the building appeared little more than a small booth with a storage area above, but beside the shop a narrow breezeway through the entire structure connected street and ward interior. The *machiya* had an open facade that allowed unmediated access from the street into the shop, and from which shopkeepers often pushed their wares into the street. In addition to such regular functions, the front rooms of the structure were brought into public use during festivals and funerals. The *machiya* thus allowed the street to "filter through the building" into the ward. It was at once street, shop, house—and wall.[67]

Though based on a traditional Japanese conception of urban space, the "community wall" embodied an approach to making boundaries in any dense urban environment. It was general enough to evoke, at the same time, a ward in traditional Kyoto, a perimeter block in a European city, and a collection of parcels in a North American suburb. The strength of Maki's proposal lay in its invitation to think of urban walls not simply as objects enclosing things of value, but as valuable sites of exchange in themselves.

Many other designers explored similar ideas in the years after Maki's *Investigations*. One of these was the English-Swedish architect Ralph Erskine. In 1969 Erskine was hired by the English city of Newcastle to redesign Byker, a large working-class neighborhood on the north side of the city. Home to eighteen thousand residents, Byker had been hard hit by the disappearance of the textile industry after World War II and had lower standards of housing and sanitation than the rest of the city. However, the neighborhood had tight-knit families and community structures built up over many generations. Planners in Newcastle were intent on avoiding the infamous failures of urban renewal in Britain and the United States, which had torn apart functioning communities by eradicating existing street layouts and replacing them with apartment towers and large blocks.[68]

With this negative example in mind, Erskine wanted to preserve as much of the existing social structure of Byker as possible, even as its physical structure was to be almost completely replaced. To do this, he proposed a number of unusual measures. Erskine's design for the neighborhood preserved important cultural and social landmarks like churches, pubs, and workmen's clubs, and the architect actively solicited residents' views by moving his office to Byker during the entire project. Erskine laid out an irregular street pattern and varied the height, shape, and materials of the new residences so they bore a stronger resemblance to the buildings and streets that had preceded them. There were no large blocks, and no single housing type or street layout was allowed to dominate the landscape.[69]

The most striking feature of the plan, however, was the treatment Erskine proposed for the northern edge of Byker. City planners in Newcastle had already mandated construction of a barrier along this edge to block sound from a proposed highway, but Erskine gave them far more than a sound wall. The architect proposed that the entire northern boundary of the neighborhood be marked by a curving edifice of apartments, balconies, and gardens, up to eight stories high and nearly a mile long. Not only would the wall mitigate a predicted nuisance, then: it would also provide tangible benefits to residents of the neighborhood. The "Byker Wall," as this structure came to be called, provided a large portion of the development's housing. It shielded the rest of the neighborhood, which consisted of low terrace housing, from harsh northerly winds, making it possible for residents to cultivate lush gardens along the south-facing slope below it. The timber balconies, irregular windows, and staggered facade of the wall's south side broke up what might have been an oppressive mass, and inhabitants of the wall were invited to embellish their balconies with plantings and bright paint. The wall even incorporated several older structures. Largely because of its wall, the Byker redevelopment quickly became famous, one architectural critic later going so far as to call it "the cynosure of the world."[70]

At the time of the Byker redevelopment, Erskine was already known for housing projects in the Arctic that included similar inhabited walls, designed to moderate the harsh climate. But these projects had been set in landscapes largely devoid of people. A massive enclosing wall was a very different kind of problem in a dense city with a long history of habitation. It is here that the Byker Wall is arguably less successful. Despite the fact that the development lay immediately south of a busy commercial district, Erskine's design made little attempt to use the wall as a link between the two neighborhoods. The balconies and plantings along the wall's south side were not repeated on the nearly windowless north face. While there were sound climatic reasons for this treatment, the result

35 A clement lee: gardens and "Byker Wall," Newcastle, England, about 1980. Courtesy of Colin Dixon/Arcaid/Corbis.

was an imposing barrier when seen from the north.[71] These oversights attracted considerable criticism as the development aged. "Perhaps the most important lesson to be learned from Byker," one architectural historian has written, "is that the design of a new community . . . must consider its perceived edges and boundaries at least as much as its center.

These areas should support transition into and out of the community, not prevent it."[72]

Yet it is misleading to depict the Byker Wall as the source of the problems that later beset the Byker neighborhood. Despite the criticism leveled at it, the wall continues to be a popular place to live, and recent studies have suggested that its inhabitants are the most satisfied of Byker's residents.[73] The wall is still almost fully occupied and has remained a stable part of the neighborhood even as the low-rise housing below it has been partly abandoned. Originally the wall improved both social and natural ecology in its shadow. It produced a microclimate of lush vegetation on its southern facade and in the neighborhoods below it, and its extensive windows and balconies encouraged what urbanist Jane Jacobs called "eyes on the street."[74] It is not incidental that the low-rise housing clusters immediately adjacent to the wall remained the safest parts of Byker long after other parts of the development had fallen into disrepair.[75] It is still possible to see the Byker Wall not as the project's failing, but as its principal success, brought low not by a fundamental flaw in design, but by the Thatcherite disinvestment of the 1980s.[76]

The Byker redevelopment was a case where an inhabited wall enclosed an entirely new neighborhood. But a similar scenario took place in the nineteenth-century neighborhoods of many European cities. Here, even as the Byker Wall was making world headlines, a new kind of recovery was about to begin.

By the late nineteenth century, the perimeter block had become the standard form of housing in many of the growing industrial cities of Europe. Whether in Paris or Vienna, Copenhagen or Genoa, these buildings bore a striking resemblance to each other. A "front house" facing the street contained elegant apartments for the middle and upper classes, while "back houses" in the interior of the block sheltered the working class. The front house apartments went all the way through the structure and were thus well ventilated and lighted, but the back houses were notorious for their darkness, poor air circulation, and lack of sanitation.[77] In cities such as Berlin, back houses were separated from the street by up to three separate buildings, with the intervening courtyards used as garbage dumps and latrines. Single rooms in a typical back house housed up to twenty people. The result was a "breeding ground for illness and disease" where the infant mortality rate often exceeded 40 percent.[78] By the early twentieth century, Berlin had the highest population density in Europe, and its tenements had earned the nickname they still bear today: *Mietskasernen*, or "rental barracks."[79]

The urban block thus formed a lateral social geography in which the well-to-do inhabited the perimeter building, the working poor lived in the structures directly behind it, and the truly destitute took refuge in coal cellars and single rooms at the center of the block. This horizontal stratification also had a vertical dimension, wealth and status decreasing with each flight of stairs the residents had to climb. The Danish writer Tove Ditlevsen evoked this hierarchy in her novel *Barndommens Gade* (Childhood Street), the story of Ester, a young girl growing up in a Copenhagen back house in the 1930s. Here Ester meets her friend Ellen, who lives in the perimeter building:

> Ellen stands swinging the pink purse she is desperate to make Ester notice. She is wearing a brown velvet coat, white stockings, and brown rubber boots and is bursting with pride because she lives on the second floor of the front house, where there is an extra space she calls "my room"—even though it's really used as a coal bin.[80]

The distinct culture Ditlevsen wrote about was soon to change. In many cities where the bombing campaigns of World War II cleared block interiors, back houses were not replaced. Berlin is now only half as dense as it was before the war, as shown in the vast open areas at the center of many of its perimeter blocks. In Copenhagen, which avoided the destruction of war, the transformation occurred in the 1970s and 1980s, when sustained political action prompted the government to demolish many of the old back houses in the interests of public health.[81] Dank block interiors became sunny walled gardens, protected environments where trees, flowers, and vegetables could flourish and where residents could sit and talk while their children played safe from the street. The perimeter building itself became a kind of inhabited wall, one whose residents collectively manage the precinct it defines. In some cases these residents retain exclusive rights to use the block interior, but in many others interiors have been opened to serve as semipublic amenities, vastly increasing the open-space network of the city. In both cases, the perimeter buildings contain shops, restaurants, and apartments on their street side that go straight through to the garden within. The perimeter building is thus an inhabited, intensively used boundary with two completely distinct, yet equally important, faces. Neither face predominates, and each supports the social and economic life around it.

36 Hedging the center: block clearing and renovation, Copenhagen, Denmark, about 1980. Courtesy of Jens V. Nielsen/Danish Architectural Press.

The perimeter block is a form of urban enclosure common to many nineteenth-century urban neighborhoods. But in North America the most common urban pattern by far is not the dense tenement but rather the aggregation of freestanding houses on private lots. Today the limits of these parcels form the overwhelming majority of boundaries in the landscape. They are the places where recovering the wall literally comes closest to home.

Reimagining the Lot

East Los Angeles is a neighborhood of modest bungalows built by real estate developers in the 1910s and 1920s. The original inhabitants of these houses were working-class whites, but by the 1930s the neighborhood began to attract more and more Mexican migrants, who came to Los Angeles to work in the heavy industries of Southern California. Today, East Los Angeles is the largest community of Mexicans and Mexican Americans in the United States, a "Mexican city in the heart of Los Angeles."[82]

The lots of East Los Angeles were originally laid out following the principles of Downing and his epigones, their unfenced lawns sloping straight down from front door to sidewalk. However, the residents who began arriving in the 1930s had no knowledge of Downing, Weidenmann, or Scott. The shape and details of the neighborhood thus gradually began to change, creating a "new hybrid form of dwelling" that combined an Anglo-American suburb with the patterns and materials immigrants had brought from Mexico.[83]

The center of this hybrid form of dwelling is not the house, but the open area between house and street, once occupied by nothing more than grass. The name residents give to this space is itself a hybrid: *la yarda*. According to urban historian Margaret Crawford, the *yarda* combines the function of "plaza, courtyard, front yard, and street." Residents use this area as an active site for socializing, outdoor work, and trade. Unlike the pristine green lawns that characterize many residential neighborhoods in Los Angeles, the *yarda* is "infinitely flexible and always in flux. . . . It can become a lush jungle of plants or, paved over, a playground or car-repair shop. A driveway can serve as a dance floor, an outdoor hallway, or a space to display goods for sale."[84] In short, the *yarda* is a place where social life in all its dimensions unfolds in full public view.

Despite this public orientation, however, *yardas* are nearly always enclosed spaces, their street frontage defined by walls and fences made of affordable materials such as chain link, wrought iron, or masonry block. Residents use these everyday materials in surprising ways, mixing pride in Mexican traditions with larger aspirations for social, economic, and linguistic integration.[85] Each street in East Los Angeles has "a characteristic topography of fences; some are patrolled by dogs, others are hung with homemade signs advertising *nopales* or discount diapers, others support brightly colored brooms for sale."[86] These fences mark private space while staging a wide variety of exchanges with passersby. The *yarda* is thus *both* enclosed *and* public.

The *yarda* demonstrates that there is no inevitable conflict between boundaries and the common, that refuge and safety need not be purchased at the cost of disengagement from the street. The walls and fences of East Los Angeles are both "protective and inviting," functioning at once as expressions of identity, markers of property, and places of social and commercial transaction. Residents use simple materials and

37 Traffic and truck: *yarda* fence, Los Angeles, California, 1994. Courtesy of Margaret Crawford.

techniques to bring an economy of care and attentiveness to the legal boundaries of their parcels. By doing so they "construct community solidarity from the inside out, house by house, street by street."[87]

The *yardas* of East Los Angeles are only one of many instances where lot fences made of common materials are transformed from barriers into sites of social truck. Landscape architect Anne Whiston Spirn noticed a similar effect at Aspen Farms, a community garden in the poor urban neighborhood of West Philadelphia.

Aspen Farms had grown out of concerted action in the 1970s, when local residents cleared away trash on a vacant lot in the center of the neighborhood, brought in clean soil, and began planting vegetables. Over the years, the gardeners developed methods of intensive cultivation that yielded far more produce than they themselves could use. Though they donated part of their harvest to local nursing homes and homeless shelters, there was no way to give the excess directly to the most obvious recipients: the residents of the neighborhood itself. The solution to this problem was deceptively simple, and it involved a fence.

One of the first things the "pioneers" of Aspen Farms had done in the 1970s was enclose the vacant lot they had occupied with wire mesh; the reality of the neighborhood was such that an unsecured, uninhabited parcel would quickly turn into a site of vandalism and illegal dumping. Over the years this original fence had grown flimsy, and the gardeners decided to replace it with a new fence made of chain link. The galvanized posts of this new fence became the impromptu mechanism by which excess produce was distributed to the surrounding neighborhood. Spirn noticed that the gardeners had filled supermarket plastic bags with tomatoes, carrots, cucumbers, and other produce and hung them on the street side of the fence to be picked up by any passerby who wanted to take them. At the same time, gardeners with plots near the fence trained flowers through the chain link to edge the sidewalk. A wooden sign on the fence read "Deut. 24:19," an allusion to the passage in the Old Testament: "When you reap your harvest in the field, and have forgotten a sheaf in the field, you shall not go back to get it; it shall be for the stranger, the fatherless, and the widow."[88]

To a casual observer, the boundary of Aspen Farms would have seemed nothing extraordinary, a chain-link fence like hundreds of others in its vicinity, distinguished only by the plastic bags hanging from its posts. And yet the small modification the gardeners made yielded a clear public benefit. The fence became an expression of the gardeners' moral commitment to help those in need. By a subtle shift in attention and use, a fencing material associated primarily with urban decay and abandonment was turned into a source of nurture for its surroundings. And while the case of Aspen Farms is specific, it is by no means unique. The same approach can be observed across the urban landscape.

The neighborhood of Cambridgeport, near Boston, is a mix of Greek Revival houses, graying three-deckers, and four- and five-story brick apartment buildings. Most of the lots these structures stand on are separated from the street by fences of iron, wood, and chain link, some dating from the era of Weidenmann, or by hedges of varying heights and materials. The result is a pleasing jumble of color and shape that always offers up something new for the senses.

One fence stands out amid this variety. On a quiet side street rises a high, gabled construction of graying board and batten, partly concealed by long strands of English ivy growing from the inner side. The main gable, perpendicular to the sidewalk and fence line, shelters a gate of wrought iron in an elaborate pattern of leaves and branches. From the ridge of a smaller gable, a birdhouse sticks up on a pole. Loopholes of various sizes and shapes offer glimpses into a lush garden inside; one contains a mirror, reflecting the viewer's own face trapped behind iron bars. The fence is clearly the result of considerable intent and care.

This boundary belongs to composer John Harbison. On a wet May afternoon I pass under the gable, through a stand of bamboo, into the yard I have glimpsed through the loopholes. Foxgloves, tulips, and hellebores are coming up in neatly edged beds; patches of new grass are iridescent in the fading light. I wait several minutes before a man with freckles and wispy gray hair opens the door. Harbison extends his hand and leads me into a study and kitchen. The floorboards creak under my shoes. A massive fireplace occupies the center of the room, and in the corner stands an old Bösendorfer piano. He gives me a cup of tea and seats me at an oak table by a glass door I have glimpsed from the sidewalk. It is strange to suddenly experience house, garden, and fence as an insider.

Harbison takes a seat opposite me and tells me the story of the fence. In the 1970s, the previous owner of the property asked friends and acquaintances, Harbison among them, to help her connect two separate buildings that shared the same lot; in exchange she allowed them to live on the property as long as they liked. The fence was designed and built by one of these friends, a Vermont carpenter whom the original owner asked to bring his skill and artistry to the task. Harbison bought the house in the 1980s, promising his friend that he would maintain the fence in its original form. Doing so has involved considerable effort and expense. The gate and gables have sometimes been vandalized, and the entire fence has had to be reconstructed several times over the past three decades. I turn the conversation to the response the fence elicits from people outside. Do others notice it as I have? "All the time," Harbison confirms, and says that the mirror and bars provoke most delight from passersby. "I often see faces in the holes, people ask about the garden, and that starts other discussions." As if to illustrate the point, at that moment a man and a woman have paused under the central gable and are peering inside, just as I have done every time I have passed alongside this boundary. "That happens a lot." Harbison laughs and waves.[89]

Why does a resident of East Los Angeles paint "Welcome" on the front wall of his *yarda*? Why did the gardeners at Aspen Farms use the fence as a distribution system for produce? Why does Harbison repair his gate year after year despite the cost in time and money? The answer has less to do with generosity of spirit than with the presentation of the individual or collective self. Residents of East Los Angeles use walls and fences to proclaim an identity marked by "tensions between culture and personality, memory and innovation, Chicano and Mexican, Mexico and America."[90] Harbison maintains his fence because by doing so he honors a remembered friendship and attracts curiosity and admiration from others. The gardeners of Aspen Farms hung plastic bags of vegetables from their fence because this act embodied an essential part of their identity as a neighborhood organization and symbolized a moral obligation to help those in need. In each of these cases, walls are a kind of mask, a public face in the original sense of publicity: "that which is manifest and open to general observation."[91]

Yet these examples also illustrate that walls are more than just signboards made by some to be "read" by others. The mask of a wall, once donned, is unavoidably *relational*; it sets in motion a performance between insiders and outsiders, actors and audience, that neither can fully predict or control. This is why the performances at walls are often highly scripted or ritualized, as any look at history will attest. But it is not necessary to repair to the Western Wall in Jerusalem or to the medieval city to see such performances in action. From the yearly collection of produce along the fence at Aspen Farms to the daily bustle of trade along the walls of East Los Angeles, new boundary rituals are arising, and others being revived, every day.

Rebuilding Ritual

A notice addressed to residents of the Parish of St. Giles's Cripplegate in London in 1860 reads as follows:

> Sir, you are desired to meet the rest of your Parishioners on Thursday the 17th of May 1860 at 10 o'Clock in the forenoon at the Church and from thence to go to your Parish Bounds afterwards to proceed to the [house of] Isaiah Blackwall there to dine. You are desir'd as a Wellwisher to the Preservation of this Society to send by the Bearer Twelve Shillings & Sixpence bringing this ticket for your admittance. [Signed], Church Wardens. Dinner on Table at 5 o'Clock precisely.[92]

Throughout history, boundaries have always been more than objects. They are also all those collective actions performed, over and over again, to uphold and pass on the social or legal condition that the boundary creates. Irrigation canals in Egypt were not just ditches, but the ritual of surveying performed every spring after the Nile flood. The Roman *pomoerium* was not just a line of boundary stones, but the yearly race run around them during the Lupercalia. And the parish boundary in medieval England was at once a beech in a hedge, the memory of the old man turned upside down there as a boy, and the yearly procession that gave rise to that memory.

Rituals of inscribing and reinscribing boundaries were obviated by the improvements in surveying and mapping that began in the sixteenth century. Such rituals became less important when disputes could be arbitrated by repairing to courts in possession of an accurate map. This was particularly true in England, where completion of the Ordnance Survey in the late nineteenth century meant there were fewer and fewer occasions when physically walking the perimeter of a parish or town was necessary for statutory purposes. By the early twentieth century, therefore, the practice of beating the bounds had nearly died out.[93]

Nearly. For the past several decades have witnessed renewed interest in boundary beating as a collective ritual in communities across England. A wide array of parishes and villages now beat their bounds regularly; some of these communities, such as Buckland Newton in Dorset, restored the practice after more than two centuries.[94] As stated by Common Ground, an organization that works to help communities recover their boundary beating rituals, such rituals "reintroduce people to things they have forgotten and give others the confidence to explore."[95]

But beating the bounds is by no means solely an English practice. The state of New Hampshire still mandates that "the lines between the towns in this state shall be perambulated, and the marks and bounds renewed, once in every 7 years forever, by the selectmen of the towns, or by such persons as they shall in writing appoint for that purpose."[96] (Massachusetts has a similar law.) Beating the bounds is thus a secular ritual and a civic obligation, one taken seriously by communities up and down the state.[97]

While boundary beating is commonly associated with rural places, it has also been revived in urban environments where the landscape has changed nearly beyond recognition, and where old boundaries must be rediscovered from ancient maps or terriers. In 1967 the Borough of Richmond in London beat its bounds for the first time in more than forty years, since which time the borough had been absorbed into greater London and the old parish boundary turned into a brick wall in Bushy Park.[98] Similarly, every March at Leyton Marshes in the London Borough of Waltham Forest, residents beat the bounds of an area where common grazing rights were enjoyed until the nineteenth century. And every Ascension Day, in the beginning of June, the members of St. Michael's congregation in Oxford pass through the town using besoms to beat parish boundary stones—some now inside pubs and shops.

The reasons for boundary beating vary from place to place. In England the practice is still mostly organized by church parishes; it therefore becomes an event for reforging the link between the church as an institution, the inhabitants of a given place, and the physical environment. In others it is a legal obligation that retains a significant degree of public support; a recent attempt to repeal New Hampshire's boundary

38 Recovering the wall: boundary beating in Bushy Park, London, England, 1967. Courtesy of London Borough of Richmond upon Thames. Photograph by Joan Heath.

beating statute was handily defeated. Many people participate in the ritual simply because it is an ideal occasion to meet one's own neighbors, particularly when adjacent parishes or towns beat their bounds on the same day. But perhaps most commonly, beating the bounds is a way to understand more deeply the history of familiar objects; when "what had previously been an ordinary ditch [is] seen to be a boundary ditch," the landscape historian W. G. Hoskins noted in the 1980s, "[it] will acquire a greater depth of meaning."[99] As the parishioners of Culmstock in Devon concluded in their account of beating the bounds in 1977 for the first time in decades: "Our parish boundary walk . . . has surely given us an insight into more than the physical limits of our Devon parish. Always near to the surface of our thoughts were the Anglo-Saxon settlers who drew the boundaries, those who were living here when Domesday was written and the thousands of farmers and craftsmen through the centuries who lived and worked in Culmstock."[100]

In all the communities where it has reemerged over the past decades, beating the bounds appears to fulfill a fundamental need increasingly unsatisfied in the modern age: an awareness of, and sense of continuity with, *place*. By walking the boundaries of their towns and villages, inhabitants forge and reforge connections to the distant past through the common medium of landscape, using only the tools of the body and senses. Beating the bounds is a way to feel landscape in the bones once again, through stumbling over fences or walls, stepping in cowpat, putting feet into ditches, or painting markers on walls. The result among those who participate in these rituals is a lasting sense of their own shared place. Beating the bounds is a way to recall that every wall, fence, hedge, and ditch in the landscape is ultimately not about separation, but about relationships.

Yet history teaches that these relationships are not always peaceful. Boundary rituals can also involve displays of power, militarism, belligerence, and violence. This is as true for newly created rituals as for ancient ones.

In July 1947, the English civil servant Cyril Radcliffe was appointed by Lord Mountbatten, last viceroy of British India, to partition the enormous colony between its Muslim and Hindu populations before its impending independence. Radcliffe was a career bureaucrat who had never before traveled to India and would never return, yet he was given only one month to make a decision that would change the lives of millions.[101]

The boundary between India and the new Muslim country of West Pakistan was to run nearly two thousand miles from the Indian Ocean to the Himalayas.[102] Its most contentious section would be in the densely populated Plain of Punjab, between the cities of Delhi and Lahore, whose population was evenly split between Hindus and Muslims. Some villages in the Punjab were dominated by a single group, others were mixed street by street and house by house. Radcliffe was given sole authority to determine, in secret, which of these settlements would remain in India and which would become part of Pakistan.

When the course of the boundary was made public on August 17, 1947, two days after Indian independence, it provoked the largest and bloodiest mass migration in history.[103] Fully one-third of the predominantly Muslim population of East Punjab, approximately 4.5 million people, fled toward Pakistan. The Hindus, Sikhs, and other non-Muslims fleeing in the opposite direction, into India, numbered over 5 million. Refugee columns were miles long, though the journey itself was almost as dangerous as remaining in place. Over the months following partition, the Plain of Punjab was riven by violence committed by both sides, as migrants on foot were subjected to sudden and brutal attacks along their entire route.[104] The confusion was heightened by the fugitive nature of the new boundary. Historian Yasmin Khan has written, "Radcliffe's judgement—which was meant to be fixed and incontestable—instead appeared soft and malleable and had little real or imagined authority behind it. People could not see the line, nor did it seem that there were enough troops available to demarcate it even if it did exist."[105] The "Radcliffe Line" sliced villages in half, cut communities off from sacred sites, agricultural land, and forests, and ignored infrastructure such as roads and railways.[106]

Most refugees going in either direction crossed the new border between India and Pakistan at only few points along its length. One of these places was Wagah, a farming village fifteen miles northeast of Lahore. Like hundreds of other settlements across the Punjab, Wagah had been split down the middle by the Radcliffe Line. In the first years after the Partition, the boundary there was marked by no gate, wall, or fence; like the new nations it separated, it was a political abstraction. One refugee who crossed at Wagah in 1947 recalled seeing only a solitary Indian flag in the middle of the plain.[107]

Today Wagah remains the one place where the border between India and Pakistan may be crossed legally.[108] The symbolic importance that derives from this status has given rise to an extraordinary ritual that draws thousands of spectators daily from across Pakistan, India, and the world. Officially titled "Raising and Lowering the Flags," the ceremony is performed each morning when the gates between the two countries are officially opened, and each evening when they are closed again. It is an elaborate display of military bluster and mock violence, performed by soldiers whose poses and uniforms are scarcely changed since the days of the British Raj. One journalist described the events this way:

> [All] stride about in sashes, cravats, and cockades like rooster combs. . . . Spectators let out joyous cheers. . . . In India, five guards march toward the gate. Long legs fold like pocketknives until their knees touch noses. . . . On each side of the national gates, one soldier lunges forward, asking his commandant's leave to approach the flag. By now both crowds are on their feet. In front, a sweating youth in an Oxford-cloth shirt shakes as he leads the cheers. "Victory for India!" he barks. The smaller Pakistani crowd pound a dent into the noise. "Pakistan! Allah Akbar!" One by one, the soldiers swagger toward their flags. . . . They snap into two facing lines. Iron-faced, a soldier from each army grasps the rope of his own flag. . . . In silent unison, the flags fall into waiting hands. . . . The guards abruptly wheel, split off, and file back to their territories. The last ones slam the gates behind them.[109]

39 Sovereign ritual: border ceremony, Wagah, India/Pakistan, about 2005. Courtesy of Michele Falzone/JAI/Corbis.

Foreign visitors in the audience are often taken aback by this strutting, staring, and slamming, which "even during peaceful times . . . is repeated every day with the same zeal, spirit and jingoism."[110] Though the performance occurs at a gate, few of the visitors who flock to the arenas on either side of the border will ever set foot on the other side. The gates themselves, flimsy steel constructions hung between unnecessarily thick columns, are little more than props for the pomp unfolding around them.

"Raising and Lowering the Flags" is not a rite of passage, then, but a staged confrontation, a symbolic battlefield re-created day after day, year after year, at the meeting place of two nations. Despite the atmosphere of violence that pervades the display, however, the ritual ultimately provides more grounds for hope than for despair. Wagah is a dramatic example of a boundary between two modern states that is expressed primarily through a collective ritual. However belligerent it may appear to outsiders, this ritual is one of the few points of communication between countries that have been frozen in hatred for the past half century, united only by their shared memories of mutual atrocity.[111] It remains the closest most of its spectators (who are also its participants) will ever come to interacting with citizens of the other country.

Wagah is thus not only a place of sublimated violence, but also a place of social exchange; that the ritual is maintained despite the other pressing social needs of India and Pakistan suggests its symbolic importance to both countries. This exchange has begun to take on new forms with a thaw in relations. Peace activists from the two countries now converge at the gate on each anniversary of the Partition, a new ritual that has prompted calls for building a "Peace Museum" at Wagah.[112] "Although a military outpost today," political scientist Syed Sikander Mehdi writes, "Wagah is evolving into a peace signpost, a junction where all the peace trains coming from different directions may converge one day."[113] It is unlikely that these changes would have occurred without the stage for interaction that the Wagah ritual created. When the next chapter is written in the history of India and Pakistan, this ritual may well turn out to have defused nationalist tensions even as it appeared to foment them.

Wagah is a reminder that boundary rituals are hardly a thing of distant places and cultures: new rituals are being created all the time. The challenge is to build such rituals not just at the borders between states, but also along walls close to home, and for purposes not of hatred, but of healing.

For many Americans, the Vietnam War was among the defining events of the twentieth century. From its beginnings in the mid-1960s to the withdrawal of American forces in 1973, over 57,000 American servicemen and servicewomen were killed or disappeared in the conflict. This number was dwarfed by the deaths of 3 million Vietnamese and Cambodians, over half of them civilians. During their time in Southeast Asia, American soldiers both witnessed and committed acts of great brutality, and many returned with physical and mental wounds that still have not healed. The Vietnam War remains the only war the United States undeniably lost, but its real legacy, for many, was the collapse of American moral authority in the world.

In 1981, eight years after the United States withdrew from Vietnam, the Vietnam Veterans Memorial Fund, a group headed by veteran Jan Scruggs, announced a national competition to construct on the National Mall in Washington, DC, a memorial to the men and women who died in the conflict. The announcement was significant because until that point the social upheaval associated with the war was still a recent memory, and many considered it too controversial for commemoration in a traditional memorial. The competition brief therefore stated that the design should "be reflective and contemplative in character" and "harmonize with its surroundings."[114] Perhaps the most unusual aspect of the brief was its stipulation that the monument include the name of every fallen or missing soldier, at that time a novel notion for a memorial of any kind.

Immediately on its announcement, the Vietnam Veterans Memorial became one of the most sought-after prizes in the world of architecture and design, and the competition attracted over fourteen hundred entries from across the globe. One of these entries was from Maya Ying Lin, a twenty-one-year-old student of architecture at Yale. Lin's design was striking in its simplicity. It called for sinking the otherwise flat surface of the Mall in a large, gently sloping chevron, as if one corner of a giant slab had compressed the ground. The wedge would be defined by two tapering granite walls at whose obtuse vertex the visitor would stand ten feet below ground level. One wall would point toward the Lincoln Memorial, the other toward the Washington Monument. Along the face of both walls would be engraved 57,692 names. As Lin described her idea in the accompanying text, the memorial would "appear . . . as a rift in the earth—a long, polished black stone wall, emerging from and receding into the earth."[115]

Lin's design won the competition unanimously, with the chairman of the eight-person jury of architects, landscape architects, and sculptors lauding its "simplicity and sense of dignity that befits an important memorial for this site."[116] The design also elicited favorable comments in the popular press, the *New York Times* calling it "a lasting and appropriate image of dignity and sadness."[117] However, Lin's proposal also provoked unprecedented opposition, with many politicians and citizens criticizing its divergence from traditional models of military heroism and remembrance. These criticisms often took on an ugly tone, with Secretary of the Interior James Watt accusing Lin of "an act of treason" and one decorated veteran famously calling the memorial "a black gash of shame and sorrow, hacked into the national visage that is the Mall."[118]

Today this criticism is all but silent. "The wall," as it has come to be known, is now one of the most important monuments in the United States, an obligatory stop in Washington for visitors and residents alike. It is the second most visited memorial in the city, attracting up to twenty thousand visitors a day. As Jan Scruggs has written: "Late at night, at dawn, someone is *always* there."[119]

The experience of the memorial is one that few forget. As you begin the procession downward along its face, the wall is little taller than your feet. Then, with each step down the gentle slope, the wall grows in height. As you approach the vertex of the two faces, it seems to enfold you in its embrace. At the bottom, the wall is a ten-foot monument whose diaphanous surface almost perfectly reflects bodies and faces through the engraved names. Here people stand, silent and reverent, pressing their hands against the surface or tracing the names of their loved ones with their fingers. Some are overcome by the experience and lean into the wall with their arms and foreheads. Like the walls of Çatalhöyük or Skara Brae, the Vietnam Veterans Memorial is a membrane between the world of the living and the world of spirits. It is the modern descendant of every numinous wall throughout history and prehistory.

But the memorial is ultimately far more than a tomb or a gravestone. It is a place of congress among the living. From its opening day, the wall has functioned as one of the prime gathering sites in Washington, a place of public celebration and remembrance. The sunken wedge-shaped form of the memorial is an almost perfect container for congregations large and small, which occur almost daily. The wall is the one place in the United States where people collectively confront the legacy of the Vietnam War. But the most common rituals associated with the wall are not public, but private and personal. Like the Western Wall in Jerusalem, the wall is a site of pilgrimage to which relatives and friends travel from near and far. Most come simply for silent contemplation, standing before the names of their friends or relatives. Others carefully impress names onto paper or aluminum foil, to hang on living room walls across the country. But perhaps most important are the things people *leave* at the wall. As in Jerusalem, the Vietnam Veterans Memorial has become a site of ritual offerings, objects, and messages. Visitors affix notes, photographs, flowers, letters, poems, and artwork to the surface of the wall, tuck them into the joints between granite slabs, or set them at the base of the wall below their loved ones' names. And this list is only a small part of the things that are left at the wall. As historian Kristin Ann Hass has written, "The Vietnam Veterans Memorial may be the only truly 'living,' national memorial in the United States. It is alive because it is transformed every day by medals and tennis balls and cans of beer left at its base."[120]

—

Standing with hundreds of pilgrims before the Western Wall in Jerusalem today, it might be tempting to think that the numinous quality of such an edifice has been altogether lost in the modern world. The Vietnam Veterans Memorial gives the lie to this notion. Little more than thirty years old, the wall shows that, even in the modern landscape,

40 Birth of a sacrament: opening day, Vietnam Veterans Memorial, Washington, DC, 1982. Courtesy of Wally McNamee/Corbis.

walls can still have resonance that reaches back to Skara Brae, the temple of Ur, and the very first gardens in the desert. But this is not all it shows. The wall also demonstrates that such resonances can and do arise from human intent and concerted action. It shows that it is possible to recover *by design* that fundamental richness of purpose that was lost when walls were reduced to mere expressions of sovereignty and property. The Vietnam Veterans Memorial is living proof that walls can yet be nurture, message, belief, dwelling—and ritual. By doing so, it challenges all the mute walls of the world to speak again.

FIVE

Toward an Ethics

In the summer of 2006, as the political debate over the construction of a fence along the border with Mexico was reaching its height, the editors of the *New York Times* solicited from prominent designers in the United States and Mexico proposals for a "national wall." The editors were assuming—correctly, as it turned out—that the border fence would be built, and that it was important to begin a discussion about the form it would take. They provided no instructions other than to think in new ways about how a wall dividing two countries might look and perform, with a view toward crafting "solutions that defy ugly problems [and] create appeal where there might be none."[1]

Some of the designers the *Times* approached refused to participate, one noting that design of walls was best left to "security and engineers." Others alluded to their personal ambivalence about the exercise. Among these was Enrique Norten, an architect from Mexico City who proposed a system of "infrastructure and connectivity that would allow our two countries to get closer." Norten's drawing showed multiple elevated highways spanning the national boundary. Los Angeles–based designer Eric Owen Moss suggested a giant linear earthwork covered with illuminated glass tubes that would "give [the boundary] a prominence over a distance." And architect Antoine Predock proposed a rampart of "tilted dirt [that] would be pushed into place by Mexican day laborers" and vanish like a mirage in the desert heat.

But perhaps the most provocative proposal was submitted by landscape architect James Corner. Rather than try to disguise the "fortified condition" of the boundary, Corner suggested that the border fence should take on "all the accoutrements of power and fortification and surveillance" that the term "national wall" implied. His drawing of this fortification—a massive sloping wall of mirrored glass—resembled a Renaissance bulwark more than a boundary between two modern states. But sheer mass was not all there was to Corner's proposal. He also suggested "turn[ing] the whole thing around to see if this new structure could have a benevolent and positive aspect." The wall not only would mark the meeting of two nations but would become a "solar power energy production line" for the United States and Mexico, as well as an entrepôt for goods traveling back and forth across the border. Closer inspection revealed ranks of long-haul trucks at the wall's base and photovoltaic panels arrayed along its surface.

Corner's design attempted to recover aspects of walls that were almost entirely lost in the debate over the border fence, which focused only on its capacity to separate the United States and Mexico. The new boundary he envisioned called into question the very idea of discrete nations in an age of international flows of people, goods, and energy. The wall would be a site of exchange in the form of commercial activity, nurture in the form of energy production, and dwelling in the sense that such a monumental installation, like Hadrian's Wall and the Great Wall of China before it, could not be maintained without a permanent human presence. The wall was not a fortification but an ecology. Though its form seemed to suggest the dominance and impregnability of a powerful nation, its prime function was to stage human congress at, and across, the border. Corner's design used a symbol of sovereignty to challenge the idea of sovereignty itself.[2]

The Limits of Property

Among the changes in almost every aspect of human thought that mark the shift to modernity, perhaps none is more significant than the notion of *natural right.* In the seventeenth and eighteenth centuries, sovereignty ceased to be seen as a matter of arbitrary power and began to be conceived as the birthright of all people. The text most closely associated with this idea of original sovereignty is Locke's *Second Treatise on Government.* But Locke did not stop with the notion that people were their own sovereigns. In one of the fateful elisions of modern times, the *Second Treatise* also posited a necessary and inevitable relation between sovereignty over the body and the sovereignty over land and objects called *property*. Because it represented the mixing of the labor of one's body with the common, property, too, was a natural right, as inalienable as the right to the body.

Many people continue to think of property in Lockean terms, as a clear and incontestable right. Yet Locke himself recognized that absolute rights of property were not workable in any real polity. Since all people have, in theory at least, equal rights to enclose parts of the common, they create civil government to balance these rights so that the rights of one do not curtail those of another. Even in Locke there is no such thing as *absolute* sovereignty of property; instead, the sovereignty of one person is limited by that of others through political agreement.

This notion has been expanded by recent scholars of property, who have challenged Locke's assumptions about its origins. They argue that property is not a single inalienable "right" at all, but rather a "bundle of rights" that are constantly being taken away, curtailed, or renegotiated in the context of larger political and social goals. As political scientist John Meyer has written, property is "a disaggregable and changeable collection rather than a unitary and near-sovereign object."[3] Lawyer Carol Rose calls property a "kind of speech, with the audience composed of all others who might be interested in claiming the object in question."[4] Urban planning scholar Harvey Jacobs has noted that notions of ownership in relation to land parcels or "sites" have "always been issues of intense social contention."[5] And in a similar vein, legal scholar Eric Freyfogle, discussing property rights in the United States, argues that claims of the inviolability of property "rest upon a poor understanding of how the law has defined landowner rights over the course of America's history. . . . Private property is made possible by law, police, and courts: it is a social institution in which public and private are necessarily joined."[6] In other words, there is no inalienable right to property that precedes law, because it is through law, custom, and communication that property comes into existence.[7] Property is the product not of naturally given sovereignty, but of politics.

This understanding of property has clear implications for the disposition of the physical environment. Writing half a century after Locke, Jean-Jacques Rousseau lamented the primal act of enclosure that created property:

> The first person who, having fenced off a plot of ground, took it into his head to say this is mine and found people simple enough to believe him, was the true founder of

civil society. What crimes, wars, murders, what miseries and horrors would the human race have been spared by someone who, uprooting the stakes or filling in the ditch, had shouted to his fellow-men: Beware of listening to this impostor; you are lost if you forget that the fruits belong to all and the earth to no one![8]

As in Locke, here the boundary of an enclosure expresses the claim to sovereignty over territory. But unlike Locke's position, it in no sense embodies a "natural right." On the contrary, in Rousseau's vision the act of making a physical boundary to establish property could be, and ought to be, *contested.*

Such contestation is all around when one begins to look for it. This is perhaps easiest to see in the case of boundaries of national, rather than individual, sovereignty. The 700-mile border fence of the United States, for example, was completed only after years of litigation and political debate. But one need look no further than the fences and walls in the everyday landscape to see how boundaries are continually constrained by politics. Communities throughout history have adopted standards and regulations for the shape, size, and location of boundaries, from the fencing rules of the medieval open-field village, to the fencing bans of nineteenth-century suburbs, to restrictions on the material composition of walls, fences, and hedges in many cities and towns today. The "right" to mark a parcel with a high hedge, a gated community with a wall, or indeed a country with eighteen-foot-high corrugated steel panels may appear absolute in theory, but in fact this "right" is subject to ongoing curtailment in the interests of wider notions of the public good.

Regulation or deliberation prevents the construction of many bad walls, though by no means all. The problem that arises is one of scale and extent. Because the modern landscape is the result of land subdivision and alienation into privately controlled parcels, it teems with walls and fences of all shapes and sizes. Most of these walls result from countless small decisions by individuals and groups. It is not feasible in this context to deliberate *every* boundary politically. Yet the mere fact that a wall is not brought into the realm of public debate does not mean it is justified. A homeowners' association in a gated community may be within its rights to build a high fence between the neighborhood and the city around it, but this does not mean that that boundary is defensible. Many actions that are *legal* within a given system of property relations may be *unjustifiable* when judged by other standards. The question is, What are those standards, and where do they come from?

More Questions to Ask a Wall

Before I built a wall I'd ask to know
What I was walling in or walling out,
And to whom I was like to give offense.

These lines from "Mending Wall" represent a common and widely accepted standard for judging the goodness or badness of a given wall. No wall that has been made without asking this question, the poem suggests, can be considered justifiable. It is not that walls should not wall people in or out, for they do this by their nature, but that they must do so in a way that consciously responds to the particular situations in which they arise.

The central message of "Mending Wall" is fundamentally *ethical*. Derived from the Greek *ethos*, "habit" or "custom," ethics is concerned not with the truth or falseness of given propositions about the world, but with the desirability of particular actions. In Aristotle's terms, ethics is that "practical wisdom" people use in order to know what constitutes good action in a given situation.[9] It is both a guide for actions that have not yet occurred—means—and a standard for judging the results—ends. Ethics is anything but abstract; it is grounded in solid things and real behavior and is practical in nature and object. In the words of philosopher Martha Nussbaum, it is a form of "complex responsiveness to the salient features of one's concrete situation."[10]

In the philosophy of both Plato and Aristotle, the primary device of any ethics is the question. It is through relentless questioning that one arrives at good action. Frost, scholar of the classics that he was, knew this. His narrator does not attempt to determine whether the wall should exist based on some abstract, universal criterion or a rational calculus about whether walls *in general* are good or bad. Such a determination, as Frost's own comments on his poem suggest, is impossible. Rather, the narrator asks, Why *this* wall when there are no cows? What purpose does *this* wall serve? Why does *this* wall make good neighbors? The ethics of the poem lies not in the answers to these questions (finally, as in any poem, there is no one answer) but in the questions themselves. The ethical difference between the two men in the poem is that the narrator asks them while his neighbor does not.

Many people who make walls today are too much like Frost's neighbor and not enough like his narrator. They are accustomed to thinking about walls within the confines of ideas about property and sovereignty that are little more than three centuries old as though these ideas were somehow "natural." Boundaries have been made by people for millennia, but the assumptions about their function and meaning have become so constricted that to build them without questioning virtually guarantees that they will act as expressions of these ideas. Like the neighbor whose unthinking dictum about good fences remains the most quoted part of "Mending Wall," people all too often engage in automatic thinking when they set about marking a boundary.

Automatic thinking is exactly what an ethics of enclosure resists. It asks questions rather than repeating proverbs. But as "Mending Wall" also shows, it is not enough merely to ask the right questions. One must pose these questions again and again, adapting them to each new situation where a wall or fence is built and fashioning new questions when the old ones no longer offer guidance. An ethics of enclosure does not end, then, with the questions of "Mending Wall." It only begins there.

Question 1. Should There Be a Wall?

This is perhaps the most fundamental question one might ask about any wall anywhere, yet it is asked far less often than it should be. It is the question that lies at the heart of "Mending Wall": Frost's narrator wants to know whether the wall should be there at all when it no longer fulfills its original function. The disappearance of that original benefit opens the possibility that the two men should allow the wall to disintegrate into nonexistence.

In some of the cases discussed in these pages, this question seems to yield a very clear answer. It is certainly possible to make an argument against the United States border fence or the Israeli "seam line obstacle" on grounds of utility: they damage political relationships, complicate travel, and increase the costs of trade. But it is equally possible to reject these walls on purely ethical grounds. One might plausibly argue that walls grounded in such a disparity of power between the people who build them and the people who experience them are simply unacceptable, whatever the details of their form, extent, or duration. This radical position shuns cost-benefit thinking and submits that certain walls in the world simply should not exist. In many cases this ethical stance can be heard in the political debates surrounding a particular wall. During the debate over the United States border fence, for example, a number of senators asked their colleagues to consider whether, on building a given wall in a given situation, a person or group is prepared to accept the possibility of a world in which others build similar walls. If the answer to this question is no, then the wall, regardless of the particular circumstances of its construction, is ethically unjustifiable.

The problem is that most of the time such clear ethical sanction is not possible. This is particularly true for most walls and fences in the urban landscape, or the boundaries of parcels rather than nations. It is indisputable that the walls and fences of gated communities degrade many urban environments, setting up an unequal relationship between people inside, who claim the right to determine the shape of that boundary, and people outside, whose lives are affected and sometimes disrupted by it. And yet it does not follow from these statements that every gated community wall is ethically unacceptable, if for no other reason than that most people likely would wish to preserve their ability to construct such walls if it were in their interest to do so. One reason it is easier to condemn national walls on ethical grounds is that their benefits, if any, are remote and unseen, while their costs are obvious and high. It is much easier to imagine the benefits that might redound from a fence built around one's own house or garden.

In most cases, then, it is not possible, or even desirable, to exclude a wall or fence on ethical grounds simply because it marks territory. But this does not mean these cases should not be subject to other forms of ethical speculation. Rather, the question whether a given wall should *exist* must yield to a more nuanced set of questions about how it *performs*. Unlike the position that a wall should not exist, such an approach implies weighing up costs and benefits: disruption to social and ecological relationships against new forms of interaction the wall might foster. And in some cases it may even mean making ethical distinctions among walls whose existence one does not condone.

Question 2. Is the Wall Contestable?

In *Discipline and Punish*, a wide-ranging history of the modern prison, Michel Foucault argued that power in the modern era has increasingly taken forms that are invisible, elusive, and insidious. Foucault connected this change to the emergence of "panopticism": "In order to be exercised . . . power had to be given the instrument of permanent, exhaustive, omnipresent surveillance, capable of making all visible, as long as it could itself remain invisible. It had to be like a faceless gaze that transformed the whole social body into a field of perception."[11]

The notion that there is an inverse relation between degrees of power and degrees of visibility is important when considering the ethical justifiability of a given wall. As the continued aversion to walls among political progressives suggests, there is a widespread tendency to equate absence of physical boundaries with openness and freedom. Yet in the modern era, lightness and invisibility have often been the handmaidens of power. In enclosure-era England, the concealment of boundaries using the device of the ha-ha naturalized a contingent set of property relations. In the American West, the capacity of barbed wire to disappear made it an ideal instrument of territorial control. And the aptly named Israeli Ministry of Defense emphasizes that the separation barrier is a "fence" rather than a "wall," part of a public relations strategy designed to make the barrier seem tenuous and impermanent. The most effective wall is, more often than not, the one that cannot be seen at all.

Ironically, then, real walls are often more contestable than their invisible counterparts. A sheer concrete wall like that of Berlin or Jerusalem is a clear and unambiguous statement of an unequal power relationship. But it is also exceedingly vulnerable to public opposition. Walls, quite simply, are easy targets. They can be painted on, things can be thrown against them, they provide instantly recognizable spaces for meeting and protesting, and as Artists Without Walls discovered, they are an ideal surface for projecting all manner of images. For this reason the sections of the separation barrier that are solid concrete, primarily in Jerusalem, have been the focus of political protest even though the wall is just one, and arguably not the most disruptive, part of an elaborate system of territorial control. The same was true in Berlin, where the concrete portion of the fortifications that imprisoned the western half of the city served as a touchstone for political resistance and artistic expression.

These two cases bring up another aspect of contestability: the lateral extent of the area on either side of the wall. A wall is contestable only to the degree that people are permitted to approach and physically interact with it. This is often as important in determining its contestability as the character of the wall itself. All "no-man's-lands," demilitarized zones, and "killing grounds" are enemies of contestation. Protesting the "seam line obstacle" outside Jerusalem is effectively impossible because of the system of trace roads and motion-sensing guns along it. The same can be observed with the United States border fence, where contestation is possible only on the Mexican side, the American side having been turned into a semi-militarized zone patrolled by border agents and now automated drones.[12] These are just two examples of a centuries-old practice that began in the Renaissance, when the sheer urban curtain walls of the Middle Ages were replaced by increasingly elaborate and extensive systems of fortification in response to the development of cannons, a transition that imposed "a dreadful social burden upon the protected population" because of their inflexibility and cost.[13] This essential change restricted not only the mutability of the wall, but also how much it could stage other forms of human interaction.

Contestability is best thought of, then, as a continuum along which various kinds of walls and fences at different scales can be arranged. In assessing this continuum, physical characteristics do not always correspond in clear and simple ways to degrees of justifiability, and transparency does not always equal benignity. It may often turn out that the most contestable parts of a wall are those that at first appear most ruthless. Determining a wall's justifiability based on its contestability will therefore often involve making uncomfortable distinctions among walls one views as unethical. It is possible to compare two equally misplaced walls—two walls that, according to the first question above, should not exist—and *still* determine that the form of one is more open to protest, and hence more ethically justifiable, than the form of the other. In other words, it is possible to construct an ethics in which the justifiability of a given wall can be measured by how far it reveals and makes explicit the particular forms of social and political control it enforces, and how far it offers a target to those who would resist that control.

Question 3. Does the Wall Foster Exchange?

No wall is impermeable, no matter how solid it may appear. Like the walls of cells, every wall in the landscape is a filter rather than a barrier. Every wall has apertures of various sizes, allowing certain things to pass and blocking the passage of others, in either direction. The difference between a cell wall and a wall in the landscape is that with the latter it is often people who control that passage. National governments set the location and hours of manned checkpoints along a border; developers decide on the location and number of gates in a fence around a planned unit development; owners choose where to place gaps in a hedge in an urban neighborhood or a rural field.

Walls can therefore help or hinder exchange, but they can never stop it entirely. The ethical question is therefore one of porosity, of the extent and type of exchange the wall allows or encourages. Such exchange must respond to the particular context in which the wall rises if it is to be deemed ethically justifiable. As with cell walls, too much porosity is not always a good in a landscape wall. For example, one reason many security forces prefer chain-link fences to concrete panels is that the former can be shot through and allow oversight of what is happening on the other side. This is a case where porosity actually *increases* sovereign control through a particular, and socially corrosive, form of exchange. By contrast, John Harbison's fence shows that a wall can be minimally permeable but still invite transactions between owner and passersby.

The term "exchange," then, is ethically neutral; it says nothing about the relative desirability of what is exchanged. This tension is highlighted by the national walls discussed here. James Corner's proposal emphasized exchange along the border in the form of trade and thus attempted to recover the vital role of walls as sites of cultural and economic interaction. The "placeless" form of this interaction in some ways suggested the universality of this impulse, freeing it from the particular conditions along the border and thus depoliticizing it. The same cannot be said of the border ceremony at Wagah, where exchange takes the form of sublimated violence; the messages sent back and forth are so particular, and so politically charged, that the experience can be disorienting for outsiders unfamiliar with the history of enmity between the two countries.

Despite their obvious differences, however, all the walls in these pages are unified by a single overarching characteristic: they are sites of intensive social production. As with the notion of contestability, one must inevitably make distinctions among particular kinds of social production when determining if a given wall is justifiable. The exchange of a "peace museum" or a demonstration using video cameras and drums is arguably more ethical than the exchange of a performance of violence. However, it is also possible to assert that any exchange, whatever form it takes, is almost always preferable to its absence. As American sociologist Lewis Coser suggested in the 1950s, in most cases, most of the time, social *conflict* is better than no *contact*.[14] Only in the most dramatic instances of violence, such as the sectarian conflict that exploded in Iraq in the years after the American invasion of 2003, might one argue that segregation is preferable to exchange.[15] It is thus necessary to construct a secondary continuum of ethical justifiability, one that considers, first, *whether* the wall fosters or prevents exchange and, second, the *kind* of exchange it stages. As with the Israeli "seam line obstacle," built in part to prevent suicide attacks, making this determination will often involve arbitrating the rights of those who build the wall and the rights of those whose lives and livelihoods are forever changed by it.

For the most part, however, the persistence of rituals along and across walls in the modern world illustrates how walls, even today, are not so much arbiters of social and political rights as "scenes for the drama of responsiveness, hospitality and responsibility."[16] In often unseen or unnoticed ways, walls and fences act as sites of all those small transactions that give depth and richness to life lived in common, particularly in urban environments. Considering the ethical justifiability of a given wall means considering how far it enables or forecloses such dramas. Along this continuum, the walls of John Harbison, Aspen Farms, and the *yardas* can be considered more "ethical" forms of enclosure. In each case, the exchange that unfolds along or near the wall is not incidental to other functions but is a central reason for the wall's existence. The owners of the wall, those who had a legal right to construct it in virtually any way they chose, nevertheless thought carefully about the kinds of exchange they wanted to stage with their surroundings, whether by using simple materials in new ways or by designing exchange into the structure of the wall itself. Such consideration need not mean erasing boundaries altogether, but rather entails thinking about the ways a given wall works itself into the fabric of life around it and subjecting the public side of the wall to the same intent and care as is afforded the territory it marks. And this suggests perhaps the most important standard by which to judge a wall or fence: its relation to the ecology of the place where it rises.

Question 4. Does the Wall Support Ecology?

The word "ecology" derives from the Greek *oikos*, "house" or "home." It thus shares a root with "economy." Coined in the late nineteenth century, this word originally designated the study of the relationships of living organisms to one another and to their physical surroundings. More recently, however, the definition of the term has grown to encompass not just the study of those relationships, but the relationships themselves. As is now widely accepted and increasingly apparent, humans are just one part of ecology, enmeshed in a thick web of relationships with the biological and physical world. To damage one part of that web is to threaten the home of the whole. An essential ethical standard when judging a given wall is thus how much it strengthens or undermines that home.

The past two centuries have seen many walls disrupt social ecology, from hedges that deprived villagers of ancient rights of access and use in the parliamentary enclosure era, to barbed wire that obliterated the economy of the Plains Indians in the 1870s, to the United States border fence that cuts through cities, towns, and Indian nations today. These are well-known cases of a physical boundary's undermining complex relationships built up over centuries between people and the places they inhabit. But what is perhaps less known about these same cases is how they also compromised natural ecology. In the American West, barbed wire was instrumental in eliminating the bison; in England, parliamentary hedges included few plant species compared with their medieval predecessors; and the American border fence has seriously compromised the habitat of endangered animal species such as the ocelot and jaguarundi and has sliced in half nature reserves owned by the Audubon Society and Nature Conservancy that contain the last two remaining stands of sabal palms in the United States.[17]

In many cases this disruption of social and natural ecology is the result of walls that obey the arbitrary logic of parcel boundaries or the boundaries of nations. Yet evidence suggests that it is also possible to strengthen ecology even in the subdivided world that most people have inherited as part of their patrimony. It has been estimated that the hedge networks of Britain are as important as national parks in providing habitat for threatened species. Because they are exposed to the sun on both sides, hedges are "probably richer in woodland edge insects than are woods."[18] Large-scale hedge networks, it is increasingly being found, are essential corridors for the movement of animals across the landscape.[19] Hedges are also important for increasing a sense of collective identity and place among those who come in contact with them. Objects that mark parcel boundaries thus arguably support human and natural ecology at least as often as they compromise it.

Sometimes, however, strengthening ecology entails either ignoring or challenging legal boundaries. Such is the case with the Great Green Wall of the Sahara, which necessarily diverges from national borders in order to improve the ecology of an entire continent (and arguably the world). This example is striking because of its sheer ambition, but similar processes happen all the time in less dramatic ways, as when two landowners agree to divert a fence or wall around an old tree. Thinking about the relation between walls and ecology means questioning the invisible structure of legal parceling within which most people live and confronting the reality that the boundaries of parcels are economic and political fictions in ways that walls and fences are not. By constructing *Running Fence*, Christo and Jeanne-Claude called attention to the fundamentally arbitrary nature of legal boundaries in the modern world. But they did so in a way that created its own, albeit brief, social ecology—one that has remained in the memories of the people who experienced it three decades ago.

One way to think about this capacity of walls and fences to support the relationship between physical boundaries and natural and social ecology is the ecotone. According to landscape ecologist Richard Forman, an ecotone is "the overlap or transition zone between two plant or animal communities." Ecotones often occur along boundaries, which "exhibit distinctive characteristics unlike the adjacent ecosystems, and [may] thus be considered a system itself."[20] The physical boundary is not simply a membrane regulating passage between areas on either side, but a distinct and invaluable ecology in itself. An example of this in the natural world is the lee of a hedge, which is inherently favorable to particular birds and insects. But ecotones do not exist only in natural ecology. As the *faubourgs* of the Middle Ages and the Byker Wall show, walls also create new zones of social activity, fundamentally *distinct* in the people they house and the activities they stage.

Conceiving of walls as areas rather than as objects requires striking a critical balance between *width* and *porosity*. If a wall is to support ecology to the maximum extent, it needs to be thick enough to support the functions associated with that ecology but thin enough, and porous enough, to foster exchange between its two sides. The combination of these two aspects, ecology and exchange, adds up to nurture.

Question 5. Does the Wall Nurture?

Walls have always created the conditions for human societies to survive and thrive. The walls of Skara Brae, the earthen banks of Cornwall, the great wall of Uruk—all were central to creating and sustaining the distinct ecology where they arose, whether a Neolithic village, an Iron Age sheepherding culture, or the urban civilization of Sumer. For millennia, walls have provided nurture in the form of shelter from the environment, or even offered shelter within themselves, as the story of Rahab illustrates.

Little has done more to degrade the nurturing capacities of walls in the modern landscape than their close association with property rights. Walls continue to nurture, but the beneficiaries of such nurture are usually the individuals or groups that hold title to the land they enclose. In this context, a wider notion of nurture is needed in the way walls are built and judged. The cases discussed here show that such a wider notion is possible even within a highly parceled environment. Often this means returning to older forms of bounding, like the medieval hedge that bore edible fruits and provided many of the basic materials necessary to life. There is no reason such offerings cannot be made in the modern landscape, as the gardeners of Aspen Farms showed with their edible fence.

Yet fruits and vegetables were never more than an accessory benefit in West Philadelphia. The real nurture the fence provided was social and even spiritual. Such spiritual sustenance is no less a part of the story of walls than are power and dominance. From the temple walls of Ur and Uruk, to the walls that fused the living and the dead at Çatalhöyük, to the Western Wall, from which the "divine Presence" is said never to depart, walls have ever been sites of numinous experience. Nor are these cases relics of the ancient past. The Vietnam Veterans Memorial, for many, is holier than any cathedral, more nurturing than any shrine. There is no reason to think the walls and fences of a subdivided landscape cannot also begin to serve such functions, as the rich topography of religious images mounted in East Los Angeles lot fences demonstrates.

Nurture is thus another standard for determining the justifiability of a given wall. It makes it possible to say, for example, that a garden hedge composed of diverse plant species that offer their fruit to passersby is in some sense more ethical than a hedge made of hawthorn and nothing else. The key to recovering these nurturing aspects of walls, both social and spiritual, is considering how well they nurture not only people in the insides they mark, but also outsiders.

In short, it is no longer ethically sufficient, if it ever was, for a wall to nurture only those it encloses, protects, enfolds. Its benefits must extend to the larger society around it. John Harbison's fence and the walls of *yardas* show that, given ongoing attention, this is possible, even simple. The nurture a wall provides will always be closely related to the care put into its making. In other words: an ethics of enclosure comes down, ultimately, to craft.

Question 6. Is the Wall Craft?

At the center of ethics as Aristotle understood it was what he called *kalon*, a word that could mean "beautiful," "noble," or "fine." For Aristotle, *kalon* had to do with the making of good artifacts: artifacts from which nothing could be taken away and to which nothing further could be added without degrading their "virtue."[21] Aristotle saw such virtue—which extended to poetry, music, and drama—as very close to ethical virtue. A well-crafted artifact or a well-crafted project were by their very definition both ethical and *kalon*. *Kalon* was thus part of an ethical standard that Aristotle saw as applying to the making and shaping of any thing.

This Aristotelian notion of craft would come to inform much thinking about the nature of art in the Renaissance; Michelangelo is said to have claimed he created the *David* by subtracting material until the perfect essence of the statue emerged. Yet there is also something static, even lifeless, in this idea. Once a thing is *kalon*, the process of craft is complete; indeed, it *must* stop, for to pursue it further would be to degrade the "virtue" of the thing made. There is little room in this vision for the messiness, mistakes, and blind alleys of human life. In short, most people, for most of their lives, cannot hope to live up to Aristotle's standard.

But there is another way to think about craft, one that has less to do with the perfection of actions or objects and more to do with their making. Sociologist Richard Sennett has described this kind of craft as "pride in one's work." In this notion of craft, the important things are circumspection, learning, reflexivity, and increasing skill. Slowness is an important corollary of this notion, since all these things, by definition, take time: "Craftsmen take pride most in skills that mature. This is why simple imitation is not a sustaining satisfaction; the skill has to evolve. The slowness of craft time serves as a source of satisfaction; practice beds in, making the skill one's own. Slow craft time also enables the world of reflection and imagination—which the push for quick results cannot."[22]

Both these visions of craft suggest what is wrong with the walls of the modern landscape. It is certainly true that in the first sense many walls fail to meet an Aristotelian standard of *kalon*. The United States border fence is in many places little more than a collection of refuse and military surplus arranged along the national boundary, and many of the walls of the urban landscape are not much better. Fences like John Harbison's stand out precisely because they are the products of careful design and craftsmanship. By contrast, most of the walls now built are mass-produced objects that fundamentally resemble each other all over the world.

But it is in the second sense of craft, as ongoing making, that the walls and fences of the modern landscape fail most seriously. Walls are now easily bought, hastily built, and quickly ignored. Such speed is the handmaid of forgetfulness, the nemesis of ethical speculation. Frost was able to think about the ethical justifiability of the wall on the edge of his field precisely because that wall *was never finished*. The very act of lifting and setting stones year after year occasioned and provided the temporal and physical space for the questioning any ethics is ultimately based on. Were the stone wall transformed into a low-maintenance chain-link fence, such questioning would likely cease.

This is why craft, in some sense, is the key to all the other ethical standards here. For it is through the very process of making and shaping actual walls that an ethics of enclosure is likely to emerge. Carefully crafted walls are prone to be contestable, since a long course of making is more likely than a short one to be public and manifest. The slowness that craft demands also fosters exchange between the builder of the wall and his surroundings, as I discovered with my own living fence. Such slowness also is likely to yield results that strengthen rather than undermine social and natural ecology. And finally, lengthy crafting makes it more probable that a given wall will nurture the particular place where it is built and the people who dwell there.

This is not a romantic vision of craft. It is a call to recover the process of ethical speculation through making things. The crucial point is that craft takes time. One must come back to the wall again and again, refining, improving, tearing down, and rebuilding. One must never take walls as finished objects; they should be seen as processes. Parcel by parcel, territory by territory, recovering the wall as craft will mean rethinking the boundary as a stage for ethical reflection, a site no longer of forgetfulness, but of care.

We Are All Outsiders

These are not the only questions one could ask of a wall. They will not be answered the same way by all people. They cannot provide flawless guidance in every case, as Aristotle knew no ethics could. Like "Mending Wall," this book is only the beginning of an ethics that will always—can only—remain unfinished. The questions it asks will always invite more nuanced questions. For example, if a wall fosters exchange, it will be desirable also to ask about the nature of that exchange and to make judgments about its character. A wall that functions as a stage for a peaceful ritual that builds the relationship between two communities is more ethical, one might argue, than a staged confrontation such as the one at Wagah. But one might equally argue that such confrontation, in almost all cases, is itself preferable to a zone, like the DMZ of Korea, that divides people and actively suppresses ongoing exchange.

Answering these questions thus will not describe most walls as *absolutely* ethical or unethical. Instead, it will help to arrange walls—both walls that exist and walls that are planned or imagined—along a continuum of justifiability. Few walls will do all the things described in this book; others will do different ones at different times. But one might postulate that the more of these questions one can answer in the affirmative, the more justifiable a given wall may be said to be. The first question—Should the wall be there?—is actually a function of the other ethical questions asked here: one cannot answer yes to it and no to all the subsequent questions. Similarly, one cannot answer no to the first question and yes to all the others. But it is not enough to ask these questions only once. Like Frost's narrator, it is necessary to return to them again and again. A wall might be porous enough for its environment one year but need to be changed the next in response to new circumstances. Boundaries, like the landscape as a whole, are dynamic, always changing, constantly in flux. Walls must have built into them the possibility, the inevitability, of that change.

This book has focused largely on thinking in new ways about the walls many people encounter every day. Most of the walls in the modern landscape are walls of property; yet there are few standards available for judging the performance of those walls that go beyond what are seen as absolute rights associated with this particular, and relatively new, form of sovereignty. Inevitably, then, recovering the wall will involve detaching the discourse about walls in the landscape from the discourse about property. We need a new attitude toward walls that recognizes them as common property. By contrast, the "ownership model" of walls, in which a single uncontested individual or group retains the right to determine the shape and character of the boundary, must be abandoned. Material boundaries impinge on the public world; thus they, like landscapes as a whole, inevitably have a moral, as well as a political, dimension.[23] However, it will not be possible to legislate such a new attitude when the economy of land and property has been built up around the entrenched myth of sovereign rights. The only way this will happen is through actual *behavior*. It is not a matter of first changing notions of property, then having property practices follow. Rather, it is only through thinking and making real things that ideas of property as absolute sovereignty will themselves, in time, be changed.[24]

An ethics of enclosure can begin to guide this thinking and making. The legacy of its absence is a landscape of bad walls, not so much because they are immoral as because they have been built unthinkingly, automatically, murmuring the same incantations as Frost's neighbor. The questions here are necessary not because they are final or complete, then, but rather because they begin to provide a common standard for what every wall, wherever it is, should do. It is on such standards that the quality of the landscape as a whole, slowly but decisively, will rise or fall.

Perhaps the ultimate standard one might adopt when building or judging a wall comes down to this: a call for concern beyond it.[25] Such a standard is built on the simple principle that for much of their lives people are more often outsiders than insiders. The modern landscape is one where nearly every territory, nearly every piece of land abuts and affects others in enormously complex ways. Most of the walls people encounter in that landscape are keeping them out, controlling their experience, and restricting their movements rather than offering protection or nurture. An ethics of enclosure recognizes this. It says that, rather than not building walls at all, it is better to expand the scope of concern to the outside, in the walls we judge and the walls we build.

In many ways this vision is similar to what French philosopher Emmanuel Lévinas called "infinite responsibility," an open-ended moral obligation to extend generosity to an unknown other.[26] But one might also argue that Lévinas's vision is not workable in practice because it fails to recognize and value our legitimate allegiance to those people and places closest to us. It fails to acknowledge that while we might extend generosity to the outside, this is an intellectual exercise; our emotions remain forever tied to the inside, to the enclosure, to the protected realm within the wall.

A better philosophical model, then, is the "veil of ignorance" described by philosopher John Rawls. According to Rawls, when constituting a political community it is necessary that each person operate from the position that "no one knows his place in society, his class position or social status, nor does any one know his fortune in the distribution of natural assets and abilities, his intelligence, strength, and the like."[27] Rawls called this stance the "original position." Only when people think about political arrangements from the original position, he argued, would any lasting form of social justice be achieved.

It is possible to perform the same mental exercise when building or judging any wall. One must ask, always, "What would my view of this wall be were I in the least advantageous position with respect to it? What benefits would it still offer?" Answering this question means adopting the Socratic position of the outsider, the stranger from another land who tries to remove her own interest from the calculus of ethical justifiability. This is a simple idea, but like many simple ideas it will be hard to implement in practice. It requires a fundamental reconception of what walls should do and what walls are. It asks people to think of walls' capacity not to separate, but to bind, not to enforce sovereignty, but to stage exchange. It requires us to think of walls not as objects that mark difference and distinction, but as the very things that bind the landscape and the people who make it up, together.

—

Our original position may have been inside the wall looking out, but that position has shifted. In the modern landscape, we are all, for much of our lives, just outside the wall, strangers passing by, glimpsing a life unfolding inside. Thinking and making from that position is one sure way to build landscapes that are more justifiable and more just.

Epilogue

I too awoke, my bed soaked with sweat, and found I'd kicked off my sheets. And no wonder: the sun shone with an intensity I'd never witnessed before, through gossamer clouds that glowed silver gray. The air was heavier and stiller than usual for an early spring morning.

Believing I'd succumbed to a sudden fever, I jumped out of bed and ran to the window. But this was hardly delirium. If anything, I felt more aware than I'd ever been, all my senses alert. The edges of objects outside had an unusual sharpness, as though a light were shining behind them, throwing them into relief and creating an almost painful contrast between a thing and what lay behind. The old wood smell of my bedroom was older and woodier, the voices of passersby more intense than yesterday yet somehow exuberant.

I shuffled into my slippers and made my way down the narrow hall as usual, toward the stairs and the first of many cups of coffee. As I passed the mirror, I glanced at my reflection. With a start I saw that my hair had turned discernibly grayer in the space between evening and morning. The skin of my neck and face sagged as if a decade had passed. But my eyes held a brilliance and clarity I hadn't seen since my youth, if ever.

I was so struck by this apparition in the mirror that I went downstairs in a trance and, forgoing my coffee, wandered out the front door dressed in nothing but my pajamas. The fresh air on my body felt so agreeable that for a moment I failed to notice that overnight early spring had turned to high summer; the trees that had been just beginning to blossom were already in full leaf. Still, there opposite was the river, nearly cresting from recent rains, just as it had been the night before. But this familiar picture wasn't enough to dispel the unease growing within me. I looked around absently and, forgetting I was half-dressed, set off down the river path into town.

I'd walked this path many times. Between it and the river's edge there now grew a great variety of tall flowering plants, from hollyhocks to lovage, and the bees and dragonflies were already hard at work. On the other side of the path, beyond my house and garden, the landscape opened up to reveal fields and pastures stretching into the distance. The path made a great sweeping arc between river and fields toward town, and I could just make out the outlines of buildings on the horizon. Today, like the objects I'd seen from my window, they stood out starkly against the silvery sky, notwithstanding the moisture in the air.

I walked along, my strange state of mind heightened by my surroundings. One of the oddest things was that so many people were around. Normally, at this early hour the path would have been almost empty, but today people of all ages were everywhere. Boys and girls in summer clothes lounged on the bank of the river, whose water, though high, was preternaturally clear. Some whooped ecstatically as they dived in and surfaced. It was as though the small strip between path and river that had been a kind of no-man's-land was suddenly the best place in the world to congregate. I approached one of the children sitting there, a freckled boy of about seventeen with impressive arm muscles and the reddest hair I'd ever seen.

"What's all the fuss today? Why are all these people here?" I asked.

"What, you don't know? It's Boundary Day."

"What day?" I asked, feeling my fever returning.

"*Boundary* Day. Today's the day we're supposed to find an orphaned boundary and adopt it."

I'd never heard a seventeen-year-old speak like this before. "Hmm," I replied, uncertain how to continue this odd conversation. "So this is the first time this festival has happened? I've been down this path many times and never noticed it."

The boy looked at me with genuine surprise. "Oh no, this has been going on for a long time, way before I was born. Our parents told us about it. That's why we're here. This is the boundary they used to claim as kids. I think it probably goes back to the first Boundary Day when it was very hot and all the kids came down to this strip between the river and the path to go swimming. Somehow after that it just happened every year. Then the kids started taking care of this land at other times too."

I looked down the narrow strip of land, remembering that yesterday it had been a rather dull grassy stretch, mowed close to the ground by overzealous municipal workers. I felt the fever returning and sat down to regain my composure. "So . . . you planted all these plants just now?" I murmured.

"Just now? Oh no!" He laughed. "These have been here for years too. We just take care of them. Actually, Boundary Day is when we *don't* take care of them, when we get to rest! All these plants are used for medicine and food, you know. There's feverfew, and coneflower, and over there is some valerian we planted last year. But here, try this one . . ."

This was surely the strangest teenager I'd ever had occasion to meet—obsessed with boundaries and expert in medicinal plants. The boy leaned down toward a small tuft of leaves growing by his dirty bare feet. He picked a large flower with five yellow petals, the likes of which I'd never seen—something akin to a giant buttercup. He handed me the flower on its stem, and I accepted it with some trepidation, inspecting the serrate edges and purplish blotches of the leaves. "This is nice . . . ," I said with some hesitation. "May I take it?"

"Take it?" he replied. "It's a wallflower. You should eat it."

After leaving the boy, with strange flower in hand I continued down the path, watching the other boys and girls along the river, some sitting, some swimming, some tending the plants the boy had told me about with such pride. I came to an old oak tree, which yesterday had been bare but today was in full leaf. I stopped under its canopy, glad to find some refuge from the mounting heat. The conversation with the boy had left me in a peculiar state. What kind of kid was this, and what was this festival I'd never heard of, despite living in this neighborhood for years? I remembered the boy's irises, piercing blue with tiny red speckles, and his redder-than-red hair. He seemed to be from another planet.

I looked down at the flower in my hand and shuddered with foreboding. *Eat this?* I thought. Its blotchy leaves and giant petals were vaguely menacing. Then I remembered how everything that morning had been new and unfamiliar. Without more thought, I bit the flower off its tall stem. Despite their soft appearance, the petals were as tough as artichokes. But the taste was unbelievably sweet, almost cloying, a sugary amalgam of blood orange and honey. Head down, I chewed away, finally swallowing the whole mass with some difficulty. *Well*, I thought. *What now?*

Just as this thought was beginning to ripen into mild panic, I heard sibilant voices in the field off to my right. I looked up at another arresting sight, a whole collection of men and women, boys and girls gathered on the edge of the field. I'd never seen so many people congregating in a farmer's field before, so I made my way over the verge into the stalks, brushing against some nettles and squinting to get a better look.

What I saw took me aback. The field edge opposite had been transformed overnight. Yesterday it was lined with barbed wire strung on posts, as it had been for as long as I could remember. Today the field was marked by an impenetrable hedge of boulders, small fruit trees, and willows. Here and there enormous beeches, oaks, and ashes jutted skyward, the songs of thrushes and bushtits emanating from their branches. I stopped in my tracks to take in this spectral boundary. It wasn't possible. . . . It looked as if the plants had been there for years, if not centuries. I made my way forward along a ditch, hearing whoops of laughter grow louder as I approached.

I walked over to the crowd and craned my neck. Over salt-and-pepper hair and balding heads I could see four full-grown men grimacing as they removed large shrubs with Weed Wrenches while others struggled to roll aside the boulders at the base of the hedge.

"What's going on here?" I asked an attractive older woman whose skin glowed like mother-of-pearl.

"It's Boundary Day!" she replied.

"Yes," I answered, happy to show that at least I knew what the day was. "But why are you going to all this trouble to make a gap in this hedge?"

"Dear, as if you didn't know! The Hole Commission warrants it!"

I felt the fever coming back. I fancied myself a political maven, but I'd never heard of any Hole Commission.

"The what?"

"Silly! The citizens' commission that determines where walls and fences should have holes in them, which holes should be opened, which ones filled, that sort of thing. You know, every Boundary Day the commission publishes its list of new holes to be made or old holes that should be filled, and groups go to carry it out."

This surely was the oddest commission I'd ever heard of. "And so you're here making a hole . . ."

"Yes, of course. It was deemed that this hedge ran a bit too long without a stile, since this path along the field edge would connect the river and the main square of town very nicely. So we're making one now."

"Ah, I see," I said hesitantly. "And if you don't mind my asking, how long has this hedge been here?"

"This hedge? Oh, decades anyway. I remember it from when I was a girl. Of course, at that time the trees in it were much smaller, and you could see the boulders better. But I remember that I'd come out here every Boundary Day with the other children and collect plums for the winter. See there, that big plum tree?"

"Yes," I answered, a knot tightening in my stomach.

"Well, I planted that tree as a girl. All this hedge was planted over time by different people, working on Boundary Day or at other times of year."

"But what about the owner of this property?" I asked. "What does he think of all this?"

"The what?" the woman asked, cocking her head as if she was hard of hearing.

"The owner. The person who owns this land."

"I'm afraid I don't know what you're talking about," she answered. "Perhaps you mean the dweller?"

"Perhaps I do," I said, trying to be cooperative. "I mean the person who decides what this boundary is going to look like."

"Ah, I see now. You mean the custodian. Yes, well, they don't control the boundaries, you know. That's done by the Boundary Standards Authority. I'm a member, actually; it's rotating." She paused as if weighing the words she'd just uttered. "You're not from around here?"

I felt less and less able to answer what should have been a simple question. "Why, yes, I am, or, I should say, I was, or . . ."

"Ah, well, never mind. You're welcome here. You look like a strapping fellow in any case, and we could certainly use your help to move that boulder over there!"

I looked at the men struggling to roll a massive granite boulder off to one side.

"Go on, then!"

Feeling suddenly a bit lighter in my soul, I made my way through the crowd, which parted to let me through, and approached one of the men.

"May I help?" I asked.

"By all means, this is a communal effort! Here, take these," he said, handing me a pair of the most beautiful tooled leather gloves I'd ever seen. I pulled them on, leaned down, and pushed.

Four boulders and twenty acquaintances later, I left the group at the hedge, made my way back to the main path, and continued toward town. As I walked I reflected that more had changed than just the seasons. The children sitting along the margin of the river, the boundary-breaking group at the hedge—all were odd enough in themselves. But little by little, even stranger sights began to appear. As I neared town, I passed an old house that had been surrounded by a board fence running alongside the path, its garden concealed by great masses of shrubbery. I'd never met the inhabitants, who so far as I could tell spent most of their time inside.

Drawing near the house on this morning, I noticed that the board fence had been replaced overnight by a most unusual boundary. It was much, much higher than the original fence—nearly fifteen feet—but was full of loopholes and openings from which projected platforms of all shapes and sizes, some as small as a birdhouse, others broad enough for two people to sit comfortably side by side. Most of the platforms were open to the air, but the largest had been enclosed to form a small room. A plant with waxy leaves and small red flowers spread across the fence's entire surface. I slowed down to examine this strange construction, wondering how it had been raised in a single evening. But so much was curious about this day that I felt less surprise than I might have an hour before when, hearing voices above my head, I noticed the inhabitants of the house, a man and woman of advancing age, perched on a generous platform at the very top of the fence.

"Pellitory," the woman said.

"Excuse me?"

"Pellitory-of-the-wall. The plant. You seemed to be admiring it."

"Ah, yes. Most handsome. But it was the wall itself I was admiring." And then, before I knew what I was saying, "Do you need help getting down?"

"Down? Heavens no! This is where we always have our breakfast."

"There, on top of the wall?"

"Wall, you say? Ah, yes, well, I suppose it might seem that way to some. We call this the boundary room."

"But . . . it's not a room at all! You're perched up there on the fence and, I must say, a risible spectacle it is! And dangerous too."

The couple looked at each other, perplexed, then down again at me. "Risible? Why so?"

"It seems a very odd place to have breakfast, that's all."

There was silence as the couple stared at me, trying to decide how to respond to this unexpected solicitude from a complete stranger. "Odd? We've been doing this all our lives, and you're the first person ever to say such a thing! It's a most natural thing to do, to have breakfast along the margin!"

"All your lives?" I queried. "But I was just passing by this place yesterday! Your house was there, sure enough, but there was no such fence, and certainly no one sitting on top of it!"

At this point the couple, who till then had seemed unsure how to answer, appeared to decide that this pajama-clad stranger was completely mad. "Well, then, I suppose you know nothing of the Margin Act either!" the woman said with a tone intended to convey that such a thing was quite impossible.

"The what?" I stammered, no doubt confirming my mental debility.

"The act that provided incentives for people to combine houses with boundaries. Everyone in this neighborhood does the same thing. Go on, you'll see!"

I was losing the capacity to be surprised by anything, whether Boundary Days or Hole Commissions or now Margin Acts. I cast my eyes down to my wrinkled hands, so much older than the day before, then back to the wall. Suddenly I was seized by a great sadness at the loss of what I'd known to be certain. Yesterday a wall was a wall and a house was a house, and they were not the same thing at all. Commissions didn't decide the location of holes. There was no such thing as a Margin Act. But here, in this strange new land, I felt my confidence in the order of things was ready to collapse just as I feared the old couple might fall from their aerie.

I looked up to say good-bye, but the two had retired to the room in the fence. I shook my head, turned, and walked on.

It had been a very strange morning, indeed. But as I approached town, it grew even stranger. I saw with mounting alarm that the couple was right. As I turned off the river path and made my way into the neighborhoods around the center of town, the landscape had been wholly transformed from the day before. Walls of all shapes and sizes, with holes and loopholes, platforms and apartments cantilevered out from them or perched on their tops, adorned streets formerly lined by chain-link

or board fences. On some I saw people reposing or eating breakfast, conversing with passersby just as the old couple had done with me. The fences were of every conceivable material: some were elaborate constructions of wood, some built from the grayish bluestone that underlies this area, some plastered as smooth and white as soap. It was at one of these that I encountered a young man, probably in his middle twenties, diligently patching the plaster where it had begun to flake off. I watched him for a moment, then complimented him on the thoroughness of his work.

"Thank you, thank you very much," he replied. "This is a particularly difficult one."

"Which one?"

"This margin, the wall here. The construction and ethics are very challenging."

I hesitated for a moment. Here was another young person speaking a language I didn't understand. "What do you mean?"

"Oh, just that the plaster is of a particularly fine type—the original custodians saw to that—but also that this a busy place, so the boundary standards are very strict."

"Ah, yes, the standards enforced by the Hole Commission."

"Ha! No, it's the other way around." He looked at me the way the old couple had done, happy to explain the obvious. "The Hole Commission is an ad hoc group of the Boundary Standards Authority, but sometimes they do act almost as an independent agency. The Authority is really the place where every aspect of boundary ethics is deliberated. They run the school, you know."

"The *school*?"

"Yes, the school for wallsmiths. Took me five years of study!"

Once again I felt the fever returning. "Five years to learn how to build a wall?"

"Yes, of course! Well, half of that was ethics anyway. Insides and outsides was the hardest course. It was difficult, learning how to think about inside as outside, and outside as inside, and all that. But it was worth it, I think. I feel I have an important job."

"Yes, most certainly you do," I affirmed, still taking in his words. "So you mean that all these walls I've been seeing in town on my walk this morning, all these were built by wallsmiths?"

"Oh, no! Wallsmiths are just master wall builders. We're in the guild. But we also act as something like educators. I teach an evening course on the technics and ethics of wallmaking, at the Betterment Center. You should stop by one evening! Are you from around here?"

Again the fateful question.

"Yes, I think. . . . At least I was at one time. But not anymore."

"Well, you're welcome. If you'll excuse me, though, I need to get back to work. But don't forget, Tuesday at seven o'clock, the Betterment Center. It's just down the road here." He gestured with his chiseled jaw. "We're always happy to have visitors."

I took my leave and continued down the street, admiring the varied walls and fences, some no doubt crafted by my recent interlocutor and his wallsmith fellows. Here were tables set through walls, there were glass cases containing items for sale or the products of this or that person's particular craft. At some walls people were gathered, heads bowed, apparently in prayer. And everywhere there were holes—big holes, small holes, loopholes, holes with glass and holes without, weep holes on the ground and peepholes at eye level. Never had I seen such a variety of "margins," as these people called them. It was this variety that made me notice one unusually monotonous parapet concealing what appeared to be a verdant garden. A large poster had been glued to it:

THE BOUNDARY STANDARDS AUTHORITY HAS DETERMINED THAT THIS WALL MAY BE IN VIOLATION OF THE MARGIN ACT. PLEASE WRITE YOUR THOUGHTS ON IT AND SUGGEST POSSIBLE CHANGES TO THE AD HOC COMMISSION ON THE PLACEMENT OF HOLES. YOUR GRAFFITI WILL BE BROUGHT BEFORE THE NEXT MEETING OF THE COMMISSION, WHICH WILL TAKE PLACE AT 6:00 PM ON AUGUST 5, 2143, AT THE GUILD HALL, 10 CORNMARKET STREET.

I felt the ground drop from beneath my feet. For some reason I looked down at my watch, which I only now realized I'd forgotten to take off the night before. It had stopped exactly at midnight.

I have no idea how long I stood staring at the notice. But I must have looked befuddled, because as my mind circled warily around the thought that I'd awakened more than a century later, I felt a hand on my shoulder. I whirled around and found myself facing a middle-aged man with a playful, mischievous face and great bushy eyebrows, traces of dirt in the crow's-feet at the corners of his eyes. He stepped back, but the smile didn't leave his eyes.

"You look like you're not feeling very well," he said, half inquiring, half asserting.

"I just . . . saw something . . . that confused me," I murmured, trying to regain my composure and remembering that I was standing in a public street in my pajamas.

"Shall we walk a bit?" he asked gently. Once again I was struck by the frankness of these people.

"Yes, thank you. . . . Though at this point I have no idea where I'm going."

"All the better!" he replied, again putting his large hand on my shoulder. "Come this way."

We walked on for some time in silence, he alternately looking straight ahead, then turning to look at me, I admiring the walls and fences that rolled past beside us. Finally, he spoke.

"You like our walls?"

His use of the possessive startled me, but I answered without hesitation. "Yes. Well, I mean, they're certainly intriguing. . . . I'm not sure I'd even call them walls, if people are living in them this way." I wanted to explain my unusual situation but demurred. "When did all this happen?"

"When did all what happen?"

"All these walls, when were they built?"

"Oh, all different times. Some are very old, though they've been repaired and changed and added to over the years. That's the whole principle behind the Margin Act, of course, that walls can't just stand there year after year. They have to be rebuilt all the time." Then, turning to me, "You're not from around here, I gather?"

"Actually, my house is just out of town, along the river path. But somehow I've never noticed these . . . constructions before."

"Well, they are distinctive, that's true enough. I came here from another place, years ago, and I can certainly say the walls there were very different indeed."

"How so?" I asked, my curiosity piqued.

"Well . . . they were . . . how should I put it? It was as if no one cared for them—they were treated as though they were the least important part of the world when they were really the most important. Funny, I don't know exactly how to put it! It's been so long since I lived in a place like that."

He paused, then resumed. "Here, let me show you something," he said, again putting his hand on my shoulder and guiding me down a side street.

We walked for some minutes until we came to the blank wall of a large building. My acquaintance stopped and turned toward it. On it was painted in red the outline of an aperture about the size of a large person. Next to the outline hung a notice: HOLE MANDATED.

"You see this?" he asked, gesturing.

"Yes, of course. But what does it mean?" I asked.

"This is where they're making a new hole in the wall. Can't you hear?"

And indeed, now that we'd stopped, I could make out dull thuds coming from inside the building.

"Yes, now I do. What about it?"

"Well, this is something that never would have happened where I come from. This wall was deemed by the Hole Commission to have a relation to the public street. The inn faces the other direction, onto a parallel street, but here the builders left the wall blank for no apparent reason. That was in the days when such things were allowed, you know. So now they're making a hole on this side so people can go in and out. That's the kind of thing the Margin Act brought about. No more blank walls, walls that do nothing, walls that only protect insides. You see?"

"So the Hole Commission deals with the walls of buildings too?"

"Yes, of course! The Margin Act applies to doors and windows too. That's the whole principle, that a boundary is a boundary, whether it's a building wall or a fence. They believe the public has an interest in where doors and windows go. I'm different; I come from a place where one person, the owner, could determine where the holes go. But here people really struggle to remember a time when apertures were the right of just one person or group."

As he said these words there was a crumbling sound and the tip of a large pickax appeared through the wall. We moved off to one side and watched the hole grow bigger.

"It's hard for them here to think about walls as anything but the center of their lives. They even bury people in their walls. They all go to rest there, because walls are a kind of religion, too. Perhaps that's why they can never really forget about walls, because they know that one day they'll all inhabit one permanently."

I'd heard too many strange stories this morning to be much concerned with this one.

"You know, they don't even call these holes doors and gates. When I happen to use those words, they find it strange that people gave them a separate name when they're really just the points in the wall where things happen. . . . But silly me," he went on after a long pause. "Here I am talking to you as though you're a stranger when you seem to understand perfectly well what I'm saying. Forgive me—I've met many people who've come from places far away where they know little about such things, and I have to explain. But shall we continue over some breakfast? I think that hole's quite big enough now, and there's food inside."

I nodded, happy enough not to continue yet another conversation leading nowhere. Yes, a stranger from a place very far away. But I had begun to doubt in the depths of my soul whether that place, that land

I'd inhabited as a native just yesterday, had ever existed at all. And, truth to tell, as this new day went on, this day that seemed so long already but was only beginning, I had less and less desire to return.

I looked behind me once more, toward the hedge, the river, and my house, nearly vanishing beyond the horizon. Then I turned to my new friend and smiled. "Yes. Let's eat."

Together, we went through.

Notes

PROLOGUE

1. Peter Marcuse, "Walls of Fear and Walls of Support," in *Architecture of Fear*, ed. Nan Ellin (New York: Princeton Architectural Press, 1997), 114.

CHAPTER ONE

1. These senators were, respectively, Jeff Sessions of Alabama, Richard Durbin of Illinois, and Patrick Leahy of Vermont. *Congressional Record—Senate*, May 17, 2006, S4653, S4661, S4663.
2. Secure Fence Act of 2006, Pub. L. No. 367, 109th Cong. (October 26, 2006).
3. Johnathan Weisman, "Border Fence Is Approved; Congress Sets Aside Immigration Overhaul in Favor of 700-Mile Barrier," *Washington Post*, October 1, 2006, sec. A, final ed.
4. The speaker was Congressman Ric Keller (R-FL). See *Congressional Record—House*, March 28, 2006, H1135.
5. "Immigration Reform Starts and Finishes with Rule of Law," *Engineering News-Record*, April 17, 2006, 60.
6. "Protect Our Borders," *Miami Herald*, July 5, 2006, sec. A.
7. See, respectively, "China Daily," *Nation* (Thailand), April 16, 2010; Phillip Coorley, "Monday Talks Fail to Break Papua Visa Deadlock," *Hobart Mercury* (Tasmania), April 10, 2006; "Albanian Lawyer Discusses Delimitation of Maritime Border Accord with Greece," *Gazeta Shqiptare* (Albania), November 8, 2010, trans. BBC Monitoring Europe, November 12, 2010; "Making S'pore More Charming and Endearing and Funny," *Straits Times* (Singapore), July 27, 2007; Michael

Djordjevich, "What to Do in Kosovo? Two-State Solution and Regional Oversight," *Washington Times*, March 20, 2007, sec. A, Op-Ed; "Finding a Way through Barriers to Peace," *Canberra Times* (Australia), September 27, 2008, sec. A; Erikka Askeland, "Good Fences Make Good Neighbours, as King Knows Well," *Scotsman* (Scotland), November 11, 2010, business ed.; Prega Govender, "President's Neighbours Bristling," *Sunday Times* (South Africa), November 7, 2010, Human Interest; "New Neighbours Are Welcome," *Montréal Gazette*, October 16, 2010, sec. B, Op-Ed, final ed.; Neil Tweedie, "When Tempers Rise above and beyond the Hedge of Reason," *Daily Telegraph* (London), September 8, 2010, Features, nat. ed., 1; Julie Bosman, "Palin Fences Off Author Neighbor," *New York Times*, May 29, 2010, sec. C, late ed.

8. See George Lakoff and Mark Johnson, *Metaphors We Live By* (Chicago: University of Chicago Press, 1980).
9. Addison Barker, "Good Fences Make Good Neighbors," *Journal of American Folklore* 64, no. 254 (1951): 421.
10. Vicesimus Knox, *Elegant Extracts, or Useful and Entertaining Passages in Prose, Selected for the Improvement of Young Persons* (1797; repr. London: J. Johnson, 1808). For the Spanish proverb, see Eleanor S. O'Kane, ed., *Refranes y frases proverbiales españolas de la Edad Media*, Anejos del Boletín de la Real Academia Española, no. 2 (Madrid: Real Academia Española, 1959), 182. For the phrase in Emerson's journal, see Edward Waldo Emerson and Waldo Emerson Forbes, eds., *Journals of Ralph Waldo Emerson*, vol. 4 (Boston: Houghton Mifflin, 1909–14), 238. The argument that Knox was the source of Emerson's journal entry is made by Ralph H. Orth in *Journals and Miscellaneous Notebooks of Ralph Waldo Emerson*, vol. 6 (Cambridge, MA: Harvard University Press, 1966), 161.
11. In Adam Winthrop et al., *Winthrop Papers*, vol. 4, *1638–1644* (Boston: Massachusetts Historical Society, 1944), 282.
12. "Fences," *Boston Cultivator*, cited in John R. Stilgoe, *Common Landscape of America, 1580 to 1845* (New Haven, CT: Yale University Press, 1982), 191.
13. Henry David Thoreau, *The Writings of Henry David Thoreau*, vol. 17, ed. Bradford Torrey (Boston: Houghton, Mifflin, 1906), 11:338, cited in George Monteiro, *Robert Frost and the New England Renaissance* (Lexington: University Press of Kentucky, 1988), 128.
14. Robert Frost, *North of Boston* (London: D. Nutt, 1914).
15. William McKenzie, "The Intersection of Immigration and Ethics," *Dallas Morning News*, August 29, 2010, Points, ed. 1.
16. John David Sweeney and James Lindroth, *The Poetry of Robert Frost* (New York: Monarch Press/Simon and Schuster, 1965), 26, cited in Lawrence Raab, "Mending Wall," in *Touchstone: American Poets on a Favorite Poem*, ed. Robert Pack and Jay Parini (Hanover, NH: University Press of New England, 1996), 204.

17. Cited in Norman Holland, *The Brain of Robert Frost: A Cognitive Approach to Literature* (London: Routledge, 1988), 26.
18. Cited in Raab, "Mending Wall," 203.
19. The poet Lawrence Raab gives particular importance to the word "between" in lines 14 and 15 ("And set the wall between us once again. / We keep the wall between us as we go."): "The repetition of *between* should give us pause and remind us of its two equally common meanings: *between* as separation, as in 'something's come between us,' and *between* as what might be shared and held in common, as in 'a secret between two people' or 'a bond between friends.' See Raab, "Mending Wall," 206.
20. Marshall McLuhan, *The Gutenberg Galaxy: The Making of Typographic Man* (Toronto: University of Toronto Press, 1962), 31.
21. See Francis Fukuyama, *The End of History and the Last Man* (London: Penguin, 1992). This teleology was highly controversial and was later substantially retracted by Fukuyama himself.
22. Gad Ya'acobi, "Tumbling Walls and Rising Hopes," *Jerusalem Post*, November 30, 1989, Op-Ed.
23. The classification is based on the annual survey of countries conducted by the nonprofit organization Freedom House. See Arch Puddington, ed., *Freedom in the World 2010: The Annual Survey of Political Rights and Civil Liberties* (New York: Freedom House, 2010).
24. Manuel Castells, *The Rise of the Network Society* (Cambridge, MA: Blackwell, 1996), 376–428.
25. Wendy Brown, *Walled States, Waning Sovereignty* (New York: Zone Books, 2010).
26. Sandro Contenta, "Journey to Limbo, by Way of Hell," *Toronto Star*, September 9, 2001, Business; Mike Davis, "The Great Wall of Capital," in *Against the Wall*, ed. Michael Sorkin (New York: Free Press, 2005), 89.
27. After the fall of the Berlin Wall in 1989, North Korea accused South Korea of building a high wall, approximately 150 miles long, along the southern edge of the Demilitarized Zone. See David E. Sanger, "Chung Hyun Ri Journal; Korean Spat: It's a Wall! No It Isn't! Is Too! Is Not!" *New York Times*, August 1, 1990, sec. A, 4.
28. Stephanie Koury, "Why This Wall?" in Sorkin, *Against the Wall*, 48–65.
29. See Oscar Newman, *Defensible Space: Crime Prevention through Urban Design* (New York: Macmillan, 1972).
30. Tom Sanchez and Robert Lang, "Security versus Status: The Two Worlds of Gated Communities," *Census Note* 2, no. 2 (Alexandria, VA: Metropolitan Institute at Virginia Tech, 2002), cited in Setha Low, *Behind the Gates: Life, Security, and the Pursuit of Happiness in Fortress America* (New York: Routledge, 2003), 15.
31. Aaron S. Rubin, "Barricades Leave Residents Divided," *Miami Herald*, June 21, 1990, final ed., 3.
32. Low, *Behind the Gates*, 16.

33. Lindsay Bremner, "Border/Skin," in Sorkin, *Against the Wall*, 132.
34. Rebecca Solnit, "Twenty-Three Steps across the Border and Back," in Sorkin, *Against the Wall*, 190.
35. Mircea Eliade, *Patterns in Comparative Religion* (1958; repr., New York: New American Library, 1974), 410.
36. Plato, *Timaeus*, 53a.
37. Jean Piaget and Bärbel Inhelder, *The Child's Conception of Space*, trans. F. J. Langdon and J. L. Lunzer (1948; repr., New York: Norton, 1967), 7.
38. Madeleine Davis, *Boundary and Space: An Introduction to the Work of D. W. Winnicott* (London: H. Karnac, 1981), 33.
39. Michael Walzer, "Liberalism and the Art of Separation," *Political Theory* 12, no. 3 (1984): 315.
40. See, for example, Will Kymlicka, "Territorial Boundaries: A Liberal Egalitarian Perspective," in *Boundaries and Justice*, ed. David Miller and Sohail Hashmi (Princeton, NJ: Princeton University Press, 2001), 249–75.
41. On the idea of boundaries and empathy, see Kok-Chor Tan, "Boundary Making and Equal Concern," *Metaphilosophy* 36, nos. 1/2 (2005): 50–67.
42. Edward Blakely and Mary Gail Snyder, *Fortress America: Gated Communities in the United States* (Cambridge, MA: Lincoln Institute of Land Policy, 1997), 177.
43. Cited in Raab, "Mending Wall," 204.
44. Aristotle, *Nicomachean Ethics* 1.7, trans. David Ross (Oxford: Oxford University Press, 1980), 11–15. See also Michael Pakaluk, *Aristotle's "Nicomachean Ethics": An Introduction* (Cambridge: Cambridge University Press, 2005), 4–6.
45. *Oxford English Dictionary*, 3rd ed., s.vv. "differentiate," "segregate."
46. *Congressional Record—Senate*, September 29, 2006, S10611–S10612.
47. Carl Schmitt, *Political Theology: Four Chapters on the Concept of Sovereignty* (Cambridge, MA: MIT Press, 1985), 5.
48. Adam Liptak, "Power to Build Border Fence Is above All U.S. Law, for Now," *New York Times*, April 8, 2008, sec. A, late ed.
49. On the route of the Israeli "Security Fence" and its deviation from the Green Line, see Eyal Weizman, "Hollow Land: The Barrier Archipelago and the Impossible Politics of Separation," in Sorkin, *Against the Wall*, 224–53.
50. "Ruling of the Supreme Court," Israeli Ministry of Defense press release of June 30, 2004, June 12, 2011, www.securityfence.mod.gov.il/pages/eng/news.htm.
51. United Nations Population Fund, *State of World Population 2007: Unleashing the Potential of Urban Growth* (New York: United Nations, 2007), 1–16.
52. Arjun Appadurai, "Deep Democracy: Urban Governmentality and the Horizon of Politics," *Environment and Urbanization* 13, no. 2 (2001): 27, cited in Mike Davis, *Planet of Slums* (London: Verso, 2006), 96.
53. On the rise in the average density at the edges of American cities, see Robert Bruegmann, *Sprawl: A Compact History* (Chicago: University of Chicago Press, 2005), 67–69.

CHAPTER TWO

1. Josh. 2:1–24, Authorized (King James) Version. References hereafter are to this version unless otherwise noted.
2. Josh. 2:1–14.
3. Josh. 2:15–24.
4. Josh. 4:13–19.
5. Josh. 6:4–5.
6. Josh. 6:21, 25.
7. Heb. 11:31: "By faith the harlot Rahab perished not with them that believed not, when she had received the spies with peace."
8. Josh. 6:21.
9. Josh. 6:26.
10. The question whether the first settlement at Jericho truly qualifies as a "city" has preoccupied urbanists for many decades. The Neolithic culture at Jericho II did not have writing, and many historians have argued that this alone disqualifies it from true urban status. They see Jericho instead as a large Neolithic village. See A. E. J. Morris, *History of Urban Form: Prehistory to the Renaissance* (New York: John Wiley, 1974), 5, 10.
11. Childe, *What Happened in History* (1942; repr., Harmondsworth, UK: Penguin Books, 1964), 58–59. See also David Lewis-Williams and David Pearce, *Inside the Neolithic Mind: Consciousness, Cosmos and the Realm of the Gods* (London: Thames and Hudson, 2005), 14.
12. Kathleen Kenyon, *Digging Up Jericho* (London: E. Benn, 1957), 67–68.
13. See Lewis-Williams and Pearce, *Inside the Neolithic Mind,* 22. The authors note that Toynbee's view is now largely discredited among archaeologists.
14. Ibid., 27.
15. Encyclopaedia Britannica Online, s.v. "Skara Brae," accessed April 3, 2013, http://www.britannica.com/EBchecked/topic/547291/Skara-Brae.
16. Gordon Childe, *Skara Brae: A Pictish Village in Orkney* (London: Kegan Paul, Trench, Trübner, 1931), 18–19.
17. Ian A. Simpson et al., "Characterizing Anthropic Settlements in North European Neolithic Settlements: An Assessment from Skara Brae, Orkney," *Geoarchaeology* 21, no. 3 (2006): 229–30.
18. Ibid., 234.
19. See, for example, Gordon Childe, *Man Makes Himself* (1936; repr., London: Watts, 1956), 82–83.
20. Many Neolithic villages were surrounded by non-defensive walls or berms. In the 1950s, archaeologist John Bradford carried out an extensive examination of Neolithic enclosures in Apulia, in the south of Italy, and discovered that the multiple concentric ditches that surrounded these villages, sometimes up to eight ditches, had "little or no defensive strength" and were primarily proprietary in nature. See John Bradford, *Ancient Landscapes* (London: Bell, 1957), 93. More recently, just a few miles from Skara Brae,

at Redlands in Orkney, archaeologists discovered a 5,000-year-old Neolithic settlement enclosed by a wall and ditch. Excavations at that site are ongoing. See Colin Richards, "Redlands Investigation," accessed April 24, 2013, http://orkneyarchaeologysociety.org.uk/index.php/features/redlands-investigtion (*sic*).

21. Childe, *What Happened in History*, 67.
22. Oliver Rackham, *The History of the Countryside* (London: J. M. Dent, 1986), 181–83.
23. Ibid., 183.
24. J. Baudry et al., "Hedgerows: An International Perspective on Their Origin, Function, and Management," *Journal of Environmental Management* 60 (2000): 9.
25. Charles Vancouver, *General View of the Agriculture of the County of Devon: With Observations on the Means of Its Improvement* (London: Phillips, 1808), cited in Ernest Pollard et al., *Hedges* (New York: Taplinger, 1974), 91.
26. A porous hedge slows wind speed up to twenty-eight hedge heights on the lee side, a decrease that translates into narrower temperature fluctuations, higher humidity, and less erosion, which all enhance soil productivity; Pollard et al., *Hedges*, 171. This is the main reason that hedges, throughout history, have been built in places with strong prevailing winds, whether England, Normandy, or central Japan. The region of Shimane, in the west of Japan, has a particularly strong tradition of using hedges to protect houses and gardens from strong prevailing winds. See Bernard Rudofsky, *Architecture without Architects* (New York: Doubleday, 1964), 132; Pollard et al., *Hedges*, 171.
27. Gerry Barnes and Tom Williamson, *Hedgerow History: Ecology, History and Landscape Character* (Macclesfield, UK: Windgather Press, 2006), 5.
28. Barnes and Williamson, *Hedgerow History*, 5; Pollard et al., *Hedges*, 109.
29. Pollard et al., *Hedges*, 191.
30. Barnes and Williamson, *Hedgerow History*, 5.
31. One of the most beloved children's stories of the twentieth century, *The Secret Garden*, tells the tale of a lonely young girl who discovers an abandoned walled garden that protects her from the "wild, low, rushing" wind of the Yorkshire moors. See Frances Hodgson Burnett, *The Secret Garden* (1911; repr., New York: Sterling, 2010), 23.
32. *The Epic of Gilgamesh*, trans. N. K. Sandars (London: Penguin Epics, 2006), vii.
33. Max Weber, *The City* (Glencoe, IL: Free Press, 1958), 78. In modern Chinese the same graph, *cheng*, can mean both "city" and "wall." See Desmond Cheung, "Chinese County Walls between the Central State and Local Society," in *Chinese Walls in Time and Space: A Multidisciplinary Perspective*, ed. Roger Des Forges et al. (Ithaca, NY: Cornell University East Asia Program, 2009), 37, 92.

34. The accomplishments of Sumerian civilization were almost unknown until the 1920s, when archaeologist Leonard Woolley excavated the ziggurat and ancient city of Ur. In addition to astonishingly advanced works of bronze and copper, Woolley also discovered that the Sumerians had used entasis, giving columns or walls a convex bulge to correct the optical illusion of concavity, two milliennia before the Greeks. See Leonard Woolley, *Ur of the Chaldees* (1935; repr., Harmondsworth, UK: Penguin, 1950).
35. Robert Sack, *Human Territoriality: Its Theory and History* (New York: Cambridge University Press, 1986), 32.
36. Lewis Mumford, *The City in History* (New York: Harcourt, Brace and World, 1961), 65.
37. Martin Brice, *Stronghold: A History of Military Architecture* (New York: Schocken Books, 1984), 34.
38. Ibid., 24.
39. Ibid., 59. Pictish bands crossed Hadrian's Wall at least five times between its completion and the withdrawal of Roman forces from the north of Britain in 383 CE.
40. Arthur Waldron, *The Great Wall of China: From History to Myth* (Cambridge: Cambridge University Press, 1990), 226, 167. An early twentieth-century observer wrote that the Great Wall, "a scant fifteen feet high . . . impressed one only by reason of the fact that it had got there at all from the sea coast." See Langdon Warner, *The Long Old Road in China* (Garden City, NY: Doubleday, Page, 1926), 63–64.
41. Woolley, *Ur of the Chaldees*, 158.
42. *Gilgamesh*, 9.
43. Cited in O. A. Dilke, *The Roman Land Surveyors: An Introduction to the Agrimensores* (Newton Abbot, UK: David and Charles, 1971), 27.
44. H. Frankfort, "Town Planning in Ancient Mesopotamia," *Town Planning Review* 21, no. 2 (1950): 102.
45. A. R. Millard, "Cartography in the Ancient Near East," in *The History of Cartography*, vol. 1, *Cartography in Prehistoric, Ancient, and Medieval Europe and the Mediterranean*, ed. J. B. Harley and David Woodward (Chicago: University of Chicago Press, 1987), 110.
46. A. F. Shore, "Egyptian Cartography," in *The History of Cartography*, vol. 1, *Cartography in Prehistoric, Ancient, and Medieval Europe and the Mediterranean*, ed. J. B. Harley and David Woodward (Chicago: University of Chicago Press, 1987), 124–25.
47. L. W. King, ed., *Babylonian Boundary Stones and Memorial Tablets in the British Museum* (London: Trustees of the British Museum, 1912), 7, cited in Joseph Rykwert, *The Idea of a Town: The Anthropology of Urban Form in Rome, Italy, and the Ancient World* (Cambridge, MA: MIT Press, 1988), 113.
48. Virgil, *First Georgic*, in Virgil, *Georgics*, trans. David Ferry (New York: Farrar, Straus and Giroux, 2005), 11.

49. Cited in Dilke, *Roman Land Surveyors*, 100.
50. Ibid., 31.
51. Ibid., 103.
52. Cited in ibid., 99.
53. Cited in ibid., 103.
54. Mary Beard et al., *Religions of Rome*, vol. 2, *A Sourcebook* (Cambridge: Cambridge University Press, 1998), 3.
55. Rykwert, *Idea of a Town*, 115–16.
56. Dilke, *Roman Land Surveyors*, 103; see also Rykwert, *Idea of a Town*, 62.
57. James Mellaart, *Çatal Hüyük: A Neolithic Town in Anatolia* (London: Thames and Hudson, 1967), 27.
58. Ibid., 223.
59. Ibid., 20.
60. Bleda S. Düring, "Social Dimensions in the Architecture of Neolithic Çatalhöyük," *Anatolian Studies* 51 (2001): 2.
61. Mellaart, *Çatal Hüyük*, 53.
62. Ibid., 118.
63. Lewis-Williams and Pearce, *Inside the Neolithic Mind*, 120.
64. Mellaart, *Çatal Hüyük*, 133.
65. Xenophon, *The Economist*, trans. H. G. Dakyns, in *The Works of Xenophon* (New York: Macmillan, 1890–97), 4:15–19.
66. Rob Aben and Saskia de Wit, *The Enclosed Garden: History and Development of the Hortus Conclusus and Its Reintroduction into the Present-Day Urban Landscape* (Rotterdam: 010 Publishers, 1999), 10.
67. Penelope Hobhouse, *Gardens of Persia* (London: Cassell Illustrated, 2003), 8.
68. Mircea Eliade, *The Sacred and the Profane: The Nature of Religion* (New York: Harcourt, Brace, 1959).
69. Woolley, *Ur of the Chaldees*, 131.
70. Morris, *History of Urban Form*, 8.
71. Mircea Eliade, *Patterns in Comparative Religion* (1958; repr., New York: New American Library, 1974), 371.
72. Sen-Dou Chang, "The Morphology of Walled Capitals," in *The City in Late Imperial China*, ed. G. W. Skinner (Stanford, CA: Stanford University Press, 1977), 90.
73. Encyclopaedia Britannica Online, s.v. "Western Wall," accessed April 4, 2014, http://www.britannica.com/EBchecked/topic/640934/Western-Wall.
74. Mumford, *City in History*, 49, 69.
75. Millard, "Cartography in the Ancient Near East," 111.
76. Claude Lévi-Strauss, *Structural Anthropology* (1958; repr., New York: Basic Books, 1963), 109.
77. Lévi-Strauss also noted that additional rings might be added in times of population increase, but that every ring, when completed, must be composed of twenty-six huts. Lévi-Strauss, *Tristes tropiques* (1955; repr., New York: Athaneum, 1974), 219. See also Lévi-Strauss, *Structural Anthropology*, 142.

78. Lévi-Strauss, *Tristes tropiques*, 221.
79. Plato, *Laws*, 6, in *The Dialogues of Plato*, trans. Benjamin Jowett (New York: Random House, 1937), 2:538.
80. Bernard Rudofsky, *Architecture without Architects: A Short Introduction to Non-pedigreed Architecture* (New York: Doubleday, 1964), 59.
81. Spiro Kostof, *The City Assembled: The Elements of Urban Form through History* (Boston: Little, Brown, 1999), 30.
82. Ibid., 30.
83. Roger Des Forges, personal communication, April 11, 2013.
84. Indeed, the term generally used during the Ming dynasty to refer to the wall, *jiu bian zhen*, translates as "the nine border garrisons." See Waldron, *Great Wall of China*, 140. Even the modern term *chang cheng* does not contain the Chinese character for "wall" (*qiang*), but uses that for "city" (*cheng*). Thus "Great Wall," correctly translated, would be "Long City."
85. Kostof, *City Assembled*, 31.
86. Mumford, *City in History*, 62.
87. Frankfort, "Town Planning," 113.
88. Paul Wheatley, *The Pivot of the Four Quarters: A Preliminary Enquiry into the Origins and Character of the Ancient Chinese City* (Chicago: Aldine, 1971), 435.
89. Kostof, *City Assembled*, 36.
90. Denis Cosgrove, *Social Formation and Symbolic Landscape*, 2nd ed. (Madison: University of Wisconsin Press, 1998), 76–77, 101.
91. Kostof, *City Assembled*, 37.
92. Ibid., 36.
93. "Life of Romulus," in *The Lives of the Noble Grecians and Romans*, trans. John Dryden (New York: Modern Library, 1979), 30–31.
94. Rykwert, *Idea of a Town*, 68.
95. Dilke, *Roman Surveyors*, 35.
96. Sicculus Flaccus, "De condicionibus agrogrum," in *Corpus agrimensorum romanorum*, ed. C. Thulin, 1, pt. 1, "Opuscula agrimensorum veterum" (Leipzig: B. G. Teubneri, 1913), cited in Rykwert, *Idea of a Town*, 116–17.
97. Ovid, *Fasti 2*, 639–64, trans. James George Frazer (London: Heinemann, 1959), 105.
98. Agnes Kirsopp Michels, "The Topography and Interpretation of the Lupercalia," *Transactions and Proceedings of the American Philological Association*, no. 84 (1953): 57–58.
99. Rykwert, *Idea of a Town*, 91.
100. Ibid., 106, 132.
101. Plutarch, "Romulus," 21.3–8, 39.
102. Rykwert, *Idea of a Town*, 94.
103. Beard, *Religions of Rome*, 119–23.
104. P. D. A. Harvey, "Local and Regional Cartography in Medieval Europe," in *The History of Cartography*, vol. 1, *Cartography in Prehistoric, Ancient, and*

Medieval Europe and the Mediterranean, ed. J. B. Harley and David Woodward (Chicago: University of Chicago Press, 1987), 464–65.

105. *Chron. Monast. Abingdon* i. 56–9, cited in C. S. Orwin and C. S. Orwin, *The Open Fields* (Oxford: Clarendon Press, 1954), 28.
106. Henry Spelman, *Councils, Decrees, Laws, and Constitutions of the English Church* (London: Richard Badger, 1639).
107. William H. Seiler, "Land Processioning in Colonial Virginia," *William and Mary Quarterly* 6, no. 3 (1949): 418.
108. W. S. Tratman, "Beating the Bounds," *Folklore* 42, no. 3 (1931): 319.
109. The passage of the Poor Law Act of 1601, which stipulated that the churchwardens of every parish be nominated as "overseers of the poor," made it necessary to fix parish boundaries with new precision. "It is conceivable that at this stage, the work of definitely marking on the ground the boundaries of the parishes and the keeping of records actually began." See Tratman, "Beating the Bounds," 319.
110. Ibid., 320–21.
111. Dorset Records Office (n.d.), cited in Tom Greeves, *The Parish Boundary* (London: Common Ground, 1987), 9.
112. Devon Records Office, 924M-B1–8, cited in Greeves, *Parish Boundary*, 10.
113. Norfolk Records Office, PD254–71, cited in Greeves, *Parish Boundary*, 9.
114. At least one critic has suggested that Frost had the Terminalia in mind when he wrote "Mending Wall." See George Monteiro, *Robert Frost and the New England Renaissance* (Lexington: University Press of Kentucky, 1988).
115. Ken Follett, *The Pillars of the Earth* (London: Pan Books, 2007), 830–31.
116. Some *poleis* followed Plato's recommendations. The highly militarized society of Sparta forbade construction of a wall on the grounds that it would make citizens lazy and decrease the effectiveness of the fighting force.
117. F. E. Winter, *Greek Fortifications* (Toronto: University of Toronto Press, 1971), 272–73.
118. Ibid., 235.
119. Ibid., 239.
120. Dean MacCannell, "Primitive Separations," in Sorkin, *Against the Wall*, 41.
121. Winter, *Fortifications*, 234.
122. Henry Cary, trans., *Herodotus: Literally Translated from the Text of Baehr with a Geographical and General Index*, vol. 2, sec. 176 (London: G. Routledge, 1892).
123. Robert Koldewey, *The Excavations at Babylon* (London: Macmillan, 1914), 32.
124. Kostof, *City Assembled*, 47; Frankfort, "Town Planning," 102.
125. Kostof, *City Assembled*, 47.
126. Ibid., 13.
127. Morris, *History of Urban Form*, 64.
128. Henri Pirenne, *Medieval Cities: Their Origins and the Revival of Trade* (Princeton, NJ: Princeton University Press, 1925), 73–74.

129. Alfred Harvey, *The Castles and Walled Towns of England* (London: Methuen, 1911), 3.
130. James W. Thompson, *An Introduction to Medieval Europe* (1937; repr., New York: W. W. Norton, 1965), cited in Morris, *History of Urban Form*, 76.
131. Frank Stenton, *Anglo-Saxon England* (Oxford: Clarendon Press, 1943), cited in Morris, *History of Urban Form*, 77.
132. Pirenne, *Medieval Cities*, 148.
133. Morris, *History of Urban Form*, 77.
134. Mumford, *City in History*, 251.
135. Kostof, *City Assembled*, 48.
136. Geoffrey Chaucer, *The Canterbury Tales* (1400; repr., New York: D. Appleton, 1855), 477.
137. Alessandro Manzoni, *The Betrothed* (1827; repr., London: Penguin Books, 1972), 522.
138. Gideon Sjoberg, *The Preindustrial City: Past and Present* (New York: Free Press, 1965), 339–43.

CHAPTER THREE

1. David Eastwood, "Communities, Protest and Police in Early Nineteenth-Century Oxfordshire: The Enclosure of Otmoor Reconsidered," *Agricultural History Review* 44, no. 1 (1996): 37.
2. Earl of Abingdon, *Case of Otmoor*, repr. 1831, 7, cited in Eastwood, "Communities, Protest and Police," 38.
3. Ibid., 39.
4. Cited in ibid., 40.
5. Ibid., 41.
6. J. M. Neeson, *Commoners: Common Right, Enclosure and Social Change in England, 1700–1820* (Cambridge: Cambridge University Press, 1993), 158–60.
7. Hans Aarsleff, "Locke's Influence," in *The Cambridge Companion to Locke*, ed. Vere Chappell (Cambridge: Cambridge University Press, 1994), 252.
8. John Gray, *Liberalism* (Milton Keynes, UK: Open University Press, 1986), 11–14.
9. John Locke, *Second Treatise of Government* (1690; repr., Indianapolis: Hackett, 1980), chap. 2, sec. 4, p. 8.
10. Thomas Hobbes, *Leviathan* (1651; repr., Oxford: Oxford University Press, 1998), chap. 13, sec. 9, p. 84.
11. Locke, *Second Treatise of Government*, 2.15.13.
12. Ibid., 5.26.18.
13. Ibid., 5.26.19.
14. Ibid., 9.123.65.
15. Ibid., 5.27.19.

16. To demonstrate the difficulty of applying this explanation of the origins of property, philosopher Robert Nozick once asked whether pouring a can of soup into the sea gave one ownership over the sea. See Robert Nozick, *Anarchy, State and Utopia* (Oxford: Blackwell, 1974), 175.
17. Locke, *Second Treatise of Government*, 5.32.21.
18. Ibid., 5.38.24.
19. Ibid., 11.136.72.
20. John Meyer, "The Concept of Private Property and the Limits of the Environmental Imagination," *Political Theory* 37, no. 1 (2009): 124.
21. Richard Marens, "Returning to Rawls: Social Contracting, Social Justice, and Transcending the Limitations of Locke," *Journal of Business Ethics* 75 (2007): 66.
22. Ibid., 67.
23. Locke, *Second Treatise of Government*, 19.222.111.
24. Cressey Dymock, "A Discovery for New Divisions, or Setting Out of Lands, as to the Best Forme: Imparted in a Letter to Samuel Hartlib, Esquire," in Samuel Hartlib, *A Discoverie for Division or Setting Out of Land, as to the Best Form* (London: Printed for Richard Wodenothe in Leaden-hall-Street, 1653), iii.
25. Hartlib, *Discoverie*, iv.
26. Ibid., v.
27. Ibid.
28. Dymock, "Discovery for New Divisions," 2.
29. Ibid., 10.
30. John Dixon Hunt, *Greater Perfections: The Practice of Garden Theory* (Philadelphia: University of Pennsylvania Press, 2000), 183.
31. Arthur Young, *General View of the Agriculture of the County of Lincoln* (London: W. Bulmer for G. Nicol, 1799), cited in W. G. Hoskins, *The Making of the English Landscape* (London: Hodder and Stoughton, 1955), 159–60.
32. G. E. Mingay, *Arthur Young and His Times* (London: Macmillan, 1975), 99.
33. John Sinclair, *Memoirs of Sir John Sinclair*, 2:111, cited in Neeson, *Commoners*, 31.
34. Arthur Young, *General View of the Agriculture of Lincolnshire* (London: Printed for Sherwood, Neely, and Jones, 1813), 488, cited in Neeson, *Commoners*, 32.
35. John Chapman, "The Extent and Nature of Parliamentary Enclosure," *Agricultural History Review* 35 (1987): 28.
36. Tom Williamson, *The Transformation of Rural England: Farming and the Landscape, 1700–1870* (Exeter, UK: University of Exeter Press, 2002), 15.
37. J. R. Wordie, "The Chronology of English Enclosure, 1500–1914," *Economic History Review* 36 (1983): 486.
38. Thomas More, *Utopia*, bk. 1 (1516; repr., Baltimore: Penguin Books, 1965), 25.
39. Ian Whyte, *Transforming Fell and Valley: Landscape and Parliamentary Enclosure in North West England* (Lancaster, UK: Center for North-West Regional Studies, 2003), 10.

40. Williamson, *Transformation of Rural England*, 7.
41. J. L. Hammond et al., *The Village Labourer, 1760–1832: A Study in the Government of England before the Reform Bill* (New York: Longmans, Green, 1913).
42. Jan Crowther, *Enclosure Commissioners and Surveyors of the East Riding* (Beverley, UK: East Yorkshire Local History Society, 1986), 9.
43. Crowther, *Enclosure Commissioners*, 21.
44. Ibid., 10.
45. G. E. Mingay, *Parliamentary Enclosure in England: An Introduction to Its Causes, Incidence, and Impact* (London: Longman, 1997), 139.
46. Hoskins, *Making of the English Landscape*, 145, 156. On the development of survey instruments in England, see A. W. Richeson, *English Land Measuring to 1800: Instruments and Practices* (Cambridge, MA: MIT Press, 1965).
47. Crowther, *Enclosure Commissioners*, 11.
48. Hoskins, *Making of the English Landscape*, 145.
49. Whyte, *Transforming Fell and Valley*, 72–73.
50. Rackham, *History of the Countryside*, 4–5.
51. H. Rider Haggard, *Rural England: Being an Account of Agricultural and Social Researches Carried Out in the Years 1901 and 1902*, vol. 1 (London: Longmans, Green, 1902), 405–6, cited in Hoskins, *Making of the English Landscape*, 159.
52. Robert C. Allen, *Enclosure and the Yeoman* (Oxford: Clarendon Press, 1992), 21. In this book Allen traces the relation between earlier enclosures and the decline of the yeomen, the only non-noble freeholding class in seventeenth-century England.
53. Neeson, *Commoners*, 158–84.
54. Ibid., 21.
55. *Cursory Remarks on Inclosures, Shewing the Pernicious and Destructive Consequences of Inclosing Common Fields &c. by a Country Farmer* (1786), 22, cited in Neeson, *Commoners*, 23.
56. Neeson, *Commoners*, 158–59.
57. Thomas Rudge, *General View of the Agriculture of the County of Gloucester* (1807), 97, cited in Neeson, *Commoners*, 29.
58. John Barrell, *The Idea of Landscape and the Sense of Place: An Approach to the Poetry of John Clare* (Cambridge: Cambridge University Press, 1972), 98–106.
59. John Clare, "The Moors," in *I Am: The Selected Poetry of John Clare* (New York: Farrar, Straus and Giroux, 2003), 91.
60. Raymond Williams, *The Country and the City* (New York: Oxford University Press, 1973), 107.
61. Encyclopaedia Britannica Online, s.v. "Massachuset," accessed December 24, 2012, http://www.britannica.com/EBchecked/topic/368399/Massachuset.
62. Encyclopaedia Britannica Online, s.v. "King Philip's War," accessed December 24, 2012, http://www.britannica.com/EBchecked/topic/318529/King-Philips-War.

63. Adam Winthrop et al., *Winthrop Papers*, vol. 2 (Boston: Massachusetts Historical Society, 1931), 141.
64. John Winthrop, "A Model of Christian Charity" (1630), in *Reading the American Past: Selected Historical Documents*, ed. Michael P. Johnson (Boston: Bedford Books, 1998), 1:49.
65. Francis J. Bremer, *John Winthrop: America's Forgotten Founding Father* (Oxford: Oxford University Press, 2003), 5.
66. D. W. Meinig, *The Shaping of America*, vol. 2 (New Haven, CT: Yale University Press, 1993), 4.
67. Ibid., 432.
68. See Richard B. Bernstein, *Thomas Jefferson* (Oxford: Oxford University Press, 2003), 142.
69. Julian Boyd, ed., *Papers of Thomas Jefferson* (Princeton, NJ: Princeton University Press, 1950–86), 39:304.
70. This was also the view held by John Adams. See Eric T. Freyfogle, *The Land We Share: Private Property and the Common Good* (Washington, DC: Island Press, 2003), 53.
71. Jefferson's vision excluded slaves and women. On Jefferson's moral struggle with this contradiction, see Bernstein, *Thomas Jefferson*.
72. Papers of the Continental Congress, 176:151, cited in William D. Pattison, *Beginnings of the American Rectangular Land Survey System, 1784–1800* (Chicago: University of Chicago Press, 1957), 3.
73. Encyclopaedia Britannica Online, s.v. "Northwest Ordinances," accessed December 29, 2012, http://www.britannica.com/EBchecked/topic/420076/Northwest-Ordinances.
74. John Stilgoe, *Common Landscape of America, 1580–1845* (New Haven, CT: Yale University Press, 1982), 102–3.
75. Pattison, *Beginnings*, 45.
76. Hildegard Binder Johnson, *Order upon the Land* (New York: Oxford University Press, 1976), 37–49.
77. A. W. Richeson, *English Land Measuring to 1800: Instruments and Practices* (Cambridge, MA: MIT Press, 1965), 6.
78. Stilgoe, *Common Landscape of America*, 101–2. See also Boyd, *Papers of Thomas Jefferson*, 8:141.
79. Crowther, *Enclosure Commissioners*, 21.
80. Pattison, *Beginnings*, 51.
81. Thomas Jefferson, letter to Thomas Mann Randolph, Washington, January 23, 1801, in Boyd, *Papers of Thomas Jefferson*, 32:500.
82. Richard White, *"It's Your Misfortune and None of My Own": A New History of the American West* (Norman: University of Oklahoma Press, 1991), 138.
83. Ibid., 139.
84. Ibid., 222.
85. Stilgoe, *Common Landscape of America*, 105.
86. White, *"It's Your Misfortune,"* 139.

87. Cited in White, *"It's Your Misfortune,"* 57.
88. Ibid., 139.
89. Ibid., 140.
90. Ibid., 147.
91. Paul G. Bourcier, "'In Excellent Order': The Gentleman Farmer Views His Fences, 1790–1860," *Agricultural History* 58, no. 4 (1984): 546–64.
92. Johnson, *Order upon the Land*, 162.
93. The journalist Marc Reisner described in detail how homesteaders were systematically misled about the aridity of the West, primarily through the specious theory that "rain follows the plow." See Marc Reisner, *Cadillac Desert: The American West and Its Disappearing Water* (New York: Viking Press, 1986), 35–36.
94. White, *"It's Your Misfortune,"* 147.
95. In Iowa, for example, speculators held title to fully two-thirds of the privately owned land in 1862. Ibid., 141.
96. Ibid., 152, 140.
97. Alan Krell, *The Devil's Rope: A Cultural History of Barbed Wire* (London: Reaktion, 2002), 19.
98. Earl W. Hayter, "Barbed Wire Fencing: A Prairie Invention; Its Rise and Influence in the Western States," *Agricultural History* 13, no. 4 (1939): 191.
99. Ibid.
100. Ibid., 195.
101. Ibid.
102. Ibid., 194.
103. White, *"It's Your Misfortune,"* 344–45.
104. Hayter, "Barbed Wire Fencing," 203.
105. *Democratic Leader* (Cheyenne, Wyoming), December 10, 1885, cited in Hayter, "Barbed Wire Fencing," 203.
106. Mollie E. Moore Davis, *The Wire Cutters* (1899; repr., College Station: Texas A&M University Press, 1997), 94.
107. Hayter, "Barbed Wire Fencing," 193.
108. James Jenkins to H. M. Teller, Pratt County, Kansas, May 26, 1883, cited in Hayter, "Barbed Wire Fencing," 202.
109. Hayter, "Barbed Wire Fencing," 202.
110. Krell, *Devil's Rope*, 39.
111. Henry D. McCallum and Frances T. McCallum, *The Wire That Fenced the West* (Norman: University of Oklahoma Press, 1965), 178.
112. McCallum and McCallum, *Wire That Fenced the West*, 165.
113. White, *"It's Your Misfortune,"* 343.
114. Ibid., 345–46.
115. *Galveston Daily News*, December 13, 1885, cited in Hayter, "Barbed Wire Fencing," 204.
116. *Galveston Daily News*, November 28, 1882, cited in Hayter, "Barbed Wire Fencing," 203.

117. Hayter, "Barbed Wire Fencing," 193.
118. Reviel Netz, *Barbed Wire: An Ecology of Modernity* (Middletown, CT: Wesleyan University Press, 2004), 13.
119. Krell, *Devil's Rope*, 38.
120. Hayter, "Barbed Wire Fencing," 196.
121. White, *"It's Your Misfortune,"* 222.
122. After the American West, the second major theater where barbed wire was used was the battlefields and trenches of World War I. See Olivier Razac, *Barbed Wire: A Political History* (New York: New Press, 2002), 32–49.
123. Ellwood reported selling barbed wire to fifty-nine railroads in 1879 alone. A single company, the Western Fence Company of Chicago, constructed "thousands of miles of wire fence" along railroads directly from its own trains, which included sleeping and dining cars for hundreds of workers. It is perhaps not surprising that railroads and barbed-wire producers were close political allies in Washington. See Krell, *Devil's Rope*, 27; Hayter, "Barbed Wire Fencing," 194–95.
124. Uvedale Price, *Essays on the Picturesque* (London, 1810), 2:148; cited in Hunt, *Greater Perfections*, 44.
125. Horace Walpole, *Anecdotes of Painting in England*, vol. 4 (Strawberry Hill, Eng.: Thomas Kirgate, 1771), 137.
126. Williams, *Country and the City*, 105–6.
127. Ann Bermingham, *Landscape and Ideology: The English Rustic Tradition, 1740–1860* (Berkeley: University of California Press, 1986), 14.
128. Stephen Daniels, *Humphry Repton: Landscape Gardening and the Geography of Georgian England* (New Haven, CT: Yale University Press, 1999), 1, 77–78.
129. Humphry Repton, *Fragments on the Theory and Practice of Landscape Gardening* (1816; repr., New York: Garland, 1982), 7; italics in original.
130. Humphry Repton, *The Red Books for Brandsbury and Glenham Hall* (Washington, DC: Dumbarton Oaks Research Library and Collection, 1994).
131. Repton, *Fragments*, 194. See also Daniels, *Humphry Repton*, 52–53.
132. Humphry Repton, *Memoir*, pt. 2 (autographed draft, British Library, Add. MS. 62112), 208, cited in Daniels, *Humphry Repton*, 210–12.
133. Cited in Bermingham, *Landscape and Ideology*, 170.
134. Daniels, *Humphry Repton*, 59. For the history of early suburbanization in London, see Robert Fishman, *Bourgeois Utopias: The Rise and Fall of Suburbia* (New York: Basic Books, 1987), 18–38.
135. Repton, *Fragments*, 233.
136. Daniels, *Humphry Repton*, 64.
137. Repton, *Fragments*, 235.
138. See Fishman, *Bourgeois Utopias*, 73–102.
139. Hunt, *Greater Perfections*, 217. See Humphry Repton, *The Landscape Gardening and Landscape Architecture of the Late Humphry Repton, Esq., Being*

His Entire Works on These Subjects, ed. J. C. Loudon (London: Longman, 1840).

140. On the picturesque, see Christopher Hussey, *The Picturesque: Studies in a Point of View* (London: Cass, 1967).
141. John Claudius Loudon, *The Suburban Gardener, and Villa Companion* (London: Longman, Orme, Brown, Green, and Longmans, 1838), 483.
142. Ibid., 660.
143. Ibid., 427–28.
144. Dolores Hayden, *Building Suburbia: Green Fields and Urban Growth, 1820–2000* (New York: Vintage, 2003), 26.
145. Downing was well acquainted with the Hudson River school painters, particularly George Inness, fifteen years his junior and also from Newburgh. See Judith K. Major, *To Live in the New World: A. J. Downing and American Landscape Gardening* (Cambridge, MA: MIT Press, 1997), 91.
146. Andrew Jackson Downing, *A Treatise on the Theory and Practice of Landscape Gardening, Adapted to North America; with a View to the Improvement of Country Residences*, 6th ed. (1841; repr., New York: A. O. Moore, 1859), viii.
147. Andrew Jackson Downing, *The Architecture of Country Houses* (New York: D. Appleton, 1851), cited in Hayden, *Building Suburbia*, 34.
148. Downing, *Treatise*, 18.
149. Ibid., ix.
150. *Horticulturist* 3 (1848): 10, cited in Kenneth Jackson, *Crabgrass Frontier: The Suburbanization of the United States* (New York: Oxford University Press, 1985), 64.
151. Andrew Jackson Downing, "Our Country Cottages," *Horticulturist* 4 (1850): 539, cited in Jackson, *Crabgrass Frontier*, 65.
152. Of Le Nôtre's garden at Versailles, which failed to impress him, Downing wrote: "Almost any one may succeed in laying out and planting a garden in right lines and may give it an air of stateliness and grandeur, by costly decorations." By contrast, only the refined few could "realize and enjoy the more exquisite beauty of natural forms." See Major, *To Live in the New World*, 39.
153. John Stilgoe, *Borderland: Origins of the American Suburb, 1820–1939* (New Haven, CT: Yale University Press, 1988), 102.
154. Downing, *Treatise*, 96.
155. Ibid., 87.
156. Ibid., 295.
157. Andrew Jackson Downing, "Our Country Villages," in *Rural Essays* (New York: Leavitt & Allen, 1869), 237–38.
158. Downing, *Treatise*, 295.
159. Ibid., 295–96.
160. Ibid., 295.

161. Jacob Weidenmann, *Beautifying Country Homes: A Handbook of Landscape Gardening* (New York: Orange Judd, 1870), 17.
162. Ibid.
163. Frank J. Scott, *The Art of Beautifying Suburban Home Grounds of Small Extent* (New York: Appleton, 1870), 51.
164. Norfolk Record Office, "Records of Barnards Ltd. of Salhouse Road, Norwich, ironfounders," catalog reference BR220, accessed January 13, 2013, http://nrocat.norfolk.gov.uk.
165. Weidenmann, *Beautifying Country Homes*, 17.
166. Scott, *Beautifying Suburban Home Grounds*, 55.
167. Calvert Vaux, *Villas and Cottages* (New York: Harper, 1864), 339, cited in Jackson, *Crabgrass Frontier*, 78.
168. Jackson, *Crabgrass Frontier*, 78.
169. *Crayon* 4 (1857): 248, cited in Therese O'Malley, "The Lawn in Early American Landscape and Garden Design," in *The American Lawn*, ed. Georges Teyssot (Princeton, NJ: Princeton Architectural Press, 1999), 82.
170. Cited in Hayden, *Building Suburbia*, 59.
171. Jackson, *Crabgrass Frontier*, 76–77.
172. Hayden, *Building Suburbia*, 60.
173. Ibid., 61.
174. Ibid., 60.
175. Ibid., 67. See also Jackson, *Crabgrass Frontier*, 98.
176. Constance Guardi, Riverside Historical Museum, personal communication, April 13, 2013. Despite this ban, Olmsted himself wrote in 1869 that "I favor fences. . . . They are of great value as making emphatic the division of freehold property—the independence of the freeholder relatively to the public & to his neighbors. . . . Doing without does not seem to me to work well—i.e. I don't find it is satisfactory for [a] long time except to eccentrics." He attributed the "want of great success" at Llewellyn Park specifically to its enforced absence of fences. Frederick Law Olmsted, letter to Edward Everett Hale, October 21, 1869, in *The Papers of Frederick Law Olmsted*, vol. 6, ed. David Schuyler and Jane Turner Censer (Baltimore: Johns Hopkins University Press, 1992), 347.
177. *Our Town—Levittown* (Levittown, NY: Levittown Property Owners Association, 1952), 16, cited in Barbara M. Kelly, *Expanding the American Dream: Building and Rebuilding Levittown* (Albany: State University of New York Press, 1993), 202.
178. *Levittown—Our Town* (Levittown, NY: Levittown Property Owners Association, 1957), 14–15, cited in Kelly, *Expanding the American Dream*, 207. The by-laws of Levittown notoriously banned not only fences, but also people "not of the Caucasian race." See Hayden, *Building Suburbia*, 135.
179. On the "craze" to ban fences in early American suburbs, see Robert M. Fogelson, *Bourgeois Nightmares* (New Haven, CT: Yale University Press, 2005), 164–68.

180. Clough Williams-Ellis, ed., *Britain and the Beast* (London: J. M. Dent, 1937), xiv.
181. "Corpse" was not the only metaphor used for this type of development in the early years of the twentieth century. In the 1893 poem "La Ville," the Belgian poet Émile Verhaeren had compared the city to an octopus, a metaphor Williams-Ellis had used in an similar anthology a decade earlier, and the Scottish town planner Patrick Geddes called urbanization an "iron glacier," a metaphor later taken up by Benton MacKaye, a founding member of the Regional Planning Association of America. See Clough Williams-Ellis, *England and the Octopus* (London: G. Bles, 1928); Patrick Geddes, *Cities in Evolution* (London, 1915); Freyfogle, *Land We Share*, 40.
182. Evelyn Waugh, *Vile Bodies* (New York: Jonathan Cape and Harrison Smith, 1930), 283–84.
183. Peter Hall, *Cities in Civilization: Culture, Innovation, and Urban Order* (London: Weidenfeld & Nicolson, 1998), 82–85.
184. H. J. Massingham, "Our Inheritance from the Past," in Williams-Ellis, *Britain and the Beast*, 9.
185. Ibid.
186. Neeson, *Commoners*, 21.
187. Williams, *Country and the City*, 107.

CHAPTER FOUR

1. Alexandra Richie, *Faust's Metropolis: A History of Berlin* (London: HarperCollins, 1998), 834.
2. Ibid., 835.
3. Charles Maier, *Dissolution: The Crisis of Communism and the End of East Germany* (Princeton, NJ: Princeton University Press, 1997), 24.
4. Zentrum für Zeithistorische Forschung, *Chronik der Mauer*, accessed April 6, 2013, http://www.chronik-der-mauer.de/index.php/de/Start/Index/id/593792.
5. Brian Ladd, *The Ghosts of Berlin: Confronting German History in the Urban Landscape* (Chicago: University of Chicago Press, 1997), 26–27.
6. Ibid., 24.
7. Michelle Standley, "The Cold War Traveler: Mass Tourism in Divided Berlin, 1945–1979" (PhD diss., New York University, 2011), 37.
8. Ibid., 4.
9. Jochim Stoltenberg, "Eine neue Zukunft," *Berliner Morgenpost*, June 14, 1990, cited in Ladd, *Ghosts of Berlin*, 32.
10. Dean MacCannell, "Primitive Separations," in *Against the Wall: Israel's Barrier to Peace*, ed. Michael Sorkin (New York: New Press, 2005), 44.
11. Dean MacCannell, personal communication, June 2, 2008.
12. Open Society Archives, "The Divide: Chapter 19: Artists Without Walls," accessed June 15, 2008, http://www.osa.ceu.hu/beta/galeria/the_divide/chapter19.html.

13. Adi Louria-Hayon, "Existence and the Other: Borders of Identity in Light of the Israeli/Palestinian Conflict," *Afterimage* 34, nos. 1/2 (2006): 22.
14. MacCannell, "Primitive Separations," 44.
15. Yishai Blank, "Who Litigates over Walls," presentation at symposium "Walls: What They Make and What They Break," Bernard and Audre Rapoport Center for Human Rights and Justice, University of Texas at Austin, February 26, 2010.
16. Jesse Lerner, "Borderline Archeology," *Cabinet*, no. 13 (2004): 34.
17. Leslie Berestein, "An Openness to Border Fence: Divider Boundless as Visual Forum for Political Expression," *San Diego Union-Tribune*, November 30, 2009, sec. A, 1.
18. Hector Tobar, "Park's Goal Lost in Border War," *Los Angeles Times*, January 6, 2009, sec. B, 1, 4.
19. Rebecca Solnit, *Wanderlust: A History of Walking* (New York: Viking, 2000), 163.
20. The Ramblers, "Your Right to Walk," accessed March 21, 2013, http://www.ramblers.org.uk/go-walking/advice-for-walkers/your-access-rights.aspx.
21. Anne Whiston Spirn, *The Language of Landscape* (New Haven, CT: Yale University Press, 1998), 120.
22. Her Majesty's Government, "Countryside and Rights of Way Act," accessed January 27, 2013, http://www.legislation.gov.uk/ukpga/2000/37/contents.
23. The Ramblers, "Gating Orders," accessed January 27, 2013, http://www.ramblers.org.uk/what-we-do/ramblers-position-on/places-we-walk/gating-orders.aspx.
24. Michel de Certeau, *The Practice of Everyday Life* (Berkeley: University of California Press, 1994), 29–30.
25. Ibid., 29.
26. Berestein, "Openness to Border Fence."
27. Ernest Pollard et al., *Hedges* (New York: Taplinger, 1974), 211.
28. J. Baudry et al., "Hedgerows: An International Perspective on Their Origin, Function, and Management," *Journal of Environmental Management* 60 (2000): 7–22; Katsue Fukamachi et al., "A Comparative Study on Trees in Hedgerows in Japan and England," in *Landscape Interfaces: Cultural Heritage in Changing Landscapes*, ed. H. Palang and G. Fry (Dordrecht: Kluwer Academic Publishers, 2003), 53.
29. Baudry, "Hedgerows," 13.
30. Pollard et al., *Hedges*, 119.
31. Ibid.
32. P. Delattre et al., "Vole Outbreaks in a Landscape Context: Evidence from a 6-Year Study of *Microtus arvalis*," *Landscape Ecology*, no. 14 (1999): 201.
33. M. Jensen, *Shelter Effect: Investigations into the Aerodynamics of Shelter and Its Effect on Climate and Crops* (Copenhagen: Danish Technical Press, 1954), 14.
34. Rackham, *History of the Countryside*, 197.
35. Royal Society for the Protection of Birds (RSPB), "Hedgerow Loss/Gain," accessed January 28, 2013, http://www.rspb.org.uk/ourwork/conservation/advice/farmhedges/loss_gain.aspx.

36. S. Oreszczyn et al., "The Meaning of Hedgerows in the English Landscape: Different Stakeholder Perspectives and the Implications for Future Hedge Management," *Journal of Environmental Management*, no. 60 (2000): 109.
37. Her Majesty's Government, "Statutory Instrument 1997 No. 1160, The Hedgerow Regulations," accessed May 22, 2008, http://www.opsi.gov.uk/si/si1997/19971160.htm.
38. RSPB, "Hedgerow Loss/Gain."
39. Baudry et al., "Hedgerows," 16.
40. Pollard et al., *Hedges*, 214, 118.
41. Richard T. T. Forman, *Land Mosaics: The Ecology of Landscapes and Regions* (Cambridge: Cambridge University Press, 1995), 195–98.
42. British Broadcasting Corporation (BBC), "Push for 'Great Green Wall of Africa' to Halt Sahara," accessed June 17, 2010, http://www.bbc.co.uk/news/10344622.
43. "The Sahel Should Already Have Been Green," *Africa News*, July 6, 2007.
44. Dan Kammen, Chief Technical Specialist for Renewable Energy and Energy Efficiency, the World Bank, accessed January 28, 2013, http://saharaforest project.com.
45. Martin Wainwright, "People's Army to Blaze a Trail along Hadrian's Wall," *Guardian*, March 12, 2010, Home Pages, 13.
46. Quoted in *Illuminating Hadrian's Wall: A Line of Light from Coast to Coast* (Hexham, Northumberland, UK: Hadrian's Wall Heritage, 2010), digital video disk.
47. Peter Brooker, "Key Words in Brecht's Theory and Practice of Theatre," in *The Cambridge Companion to Brecht*, ed. Peter Thomson and Glendyr Sacks (Cambridge: Cambridge University Press, 1994), 191.
48. On the Russian notion of *ostranenie*, "making strange," derived from Brecht's *Verfremdungseffekt*, see Viktor Shklovskii, "Art as Device," in *Theory of Prose* (Elmwood Park, IL: Dalkey Archive Press, 1990), 1–14.
49. On defamiliarization in relation to landscape, see Jacky Bowring, "To Make the Stone[s] Stony: Defamiliarisation and Andy Goldsworthy's Garden of Stones," in *Contemporary Garden Aesthetics, Creations and Interpretations*, ed. Michel Conan (Cambridge, MA: Harvard University Press, 2007), 181–98.
50. Christo and Jeanne-Claude, *Remembering the "Running Fence"* (Berkeley: University of California Press, 2010), 104.
51. Eileen Keerdoja, "A Fence to Remember," *Newsweek*, February 21, 1977, 8.
52. Christo and Jeanne-Claude, *Remembering the "Running Fence,"* 122.
53. Werner Spies, *The "Running Fence" Project: Christo* (New York: Abrams, 1977), 2.
54. Alfred Frankenstein, "An Art Critic's View of the Fence," *San Francisco Chronicle*, September 9, 1976, 26.
55. Charles Petit, "The Giant Fence Is Up and Running," *San Francisco Chronicle*, September 9, 1976, 5.
56. Keerdoja, "Fence to Remember."
57. G. Wayne Clough, "Remembrance from an Observer," in Christo and Jeanne-Claude, *Remembering the "Running Fence,"* 25.

58. Albert Elsen, "The Controversial Fence That 'Touches the Heart,'" *San Francisco Chronicle*, September 19, 1976, This World, 3.
59. Keerdoja, "Fence to Remember."
60. Cited in Brian O'Doherty, "Christo and Jeanne-Claude's *Running Fence*: Still Running," in Christo and Jeanne-Claude, *Remembering the "Running Fence,"* 63.
61. Elizabeth Broun, "Creating Joy and Beauty," in Christo and Jeanne-Claude, *Remembering the "Running Fence,"* 132.
62. Fumihiko Maki, *Investigations in Collective Form* (St. Louis: School of Architecture, Washington University, 1964), 5.
63. Ibid., 29.
64. Ibid., 27.
65. Jennifer Taylor, *The Architecture of Fumihiko Maki: Space, City, Order, and Making* (Boston: Birkhauser, 2003), 44.
66. Maki, *Investigations in Collective Form*, 67.
67. William R. Tingey, "The Principal Elements of Machiya Design," *Process: Architecture* 25 (1981): 83–88.
68. Robin Abrams, "Byker Revisited," *Built Environment* 29, no. 2 (2003): 117.
69. David Dunster, "Walled Town: Byker Redevelopment, Newcastle-upon-Tyne," *Progressive Architecture* 60, no. 8 (1979): 70.
70. Peter Davey, "Outrage," *Architectural Review*, no. 202; Arts Module 23 (1997): 1205.
71. Dunster, "Walled Town," 68.
72. Abrams, "Byker Revisited," 130.
73. Ibid., 124.
74. Dunster, "Walled Town," 68.
75. Abrams, "Byker Revisited," 124.
76. Davey, "Outrage," 1205.
77. Mumford, *City in History*, 465.
78. Richie, *Faust's Metropolis*, 164.
79. Ibid., 163.
80. Tove Ditlevsen, *Barndommens Gade* [Childhood Street] (1941; Copenhagen: Hasselbach, 1986), 14; my translation.
81. See Arne Gaardmand, *Dansk Byplanlaegning, 1938–1992* [Danish City Planning, 1938–92] (Copenhagen: Arkitektens Forlag, 1993), 234–53.
82. Margaret Crawford, "Mi casa es su casa," *Assemblage*, no. 24 (1994): 13.
83. Ibid., 14.
84. Ibid., 15.
85. Ibid., 12.
86. Ibid., 14.
87. Ibid., 12.
88. Spirn, *Language of Landscape*, 72–80.
89. John Harbison, personal communication, May 16, 2008.
90. Crawford, "Mi casa es su casa," 12.

91. Richard Sennett, *The Fall of Public Man* (New York: Alfred A. Knopf, 1977), 16.
92. W. Crouch, "To the Natives of the Parish of St. Giles's Cripplegate" (broadsheet), no. 59, 1860. Look and Learn/Peter Jackson Collection; http://www.lookandlearn.com/history-images/XJ100992/Parish-of-St-Giless-Cripplegate-beating-the-bounds-invitation-1860?img=1&search=xj100992.
93. W. S. Tratman, "Beating the Bounds," *Folklore* 42, no. 3 (1931): 321–22.
94. Tom Greeves, *The Parish Map* (London: Common Ground, 1987), 12. See also Matthew Potteiger and Jamie Purinton, *Landscape Narratives: Design Practices for Telling Stories* (New York: John Wiley, 1998), 190.
95. Cited in Potteiger and Purinton, *Landscape Narratives*, 191.
96. "Town Lines and Perambulation of Boundaries," Revised Statutes of the State of New Hampshire, Title III, sec. 51:2.
97. Christopher J. Porter, "360 Years of Perambulation," accessed April 5, 2013, http://www.nhlgc.org/publications/item_detail.asp?TCArticleID=350.
98. London Borough of Richmond upon Thames, "Community Archive: 1960s," accessed June 19, 2008, http:// www2.richmond.gov.uk/communityarchive.
99. W. G. Hoskins, *Fieldwork in Local History* (London: Faber and Faber, 1982), 40.
100. *Culmstock: A Devon Village* (n.p., n.d.), 10, cited in Tom Greeves, *The Parish Boundary* (London: Common Ground, 1987), 11.
101. Ravinder Kaur, *Since 1947: Partition Narratives among Punjabi Migrants of Delhi* (New Delhi: Oxford University Press, 2007), 70.
102. Pakistan itself was divided into West and East halves. After a bloody revolution and invasion by India in 1971, East Pakistan declared its independence and was renamed Bangladesh.
103. D. R. SarDesai, *India: The Definitive History* (Boulder, CO: Westview Press, 2008), 311.
104. Kaur, *Since 1947*, 81.
105. Yasmin Khan, *The Great Partition: The Making of India and Pakistan* (New Haven, CT: Yale University Press, 2007), 127.
106. Ibid., 126.
107. Ibid., 161.
108. Syed Sikander Mehdi, "A Peace Museum on the Wagah Border," *South Asian Journal*, no. 10 (2005): 121.
109. Claudia Kolker, "Power through Pageantry: India-Pakistan Ritual Puzzling to Outsiders," *Houston Chronicle*, January 26, 2002, sec. A, 26.
110. Mehdi, "Museum," 122.
111. Kaur, *Since 1947*, 208.
112. Ibid.
113. Mehdi, "Museum," 124.
114. Cited in Kristin Ann Hass, *Carried to the Wall: American Memory and the Vietnam Veterans Memorial* (Berkeley: University of California Press, 1998), 13.

115. Maya Lin, *Boundaries* (New York: Simon and Schuster, 2000), 4:05.
116. Cited in Jan Scruggs and Joel Swerdlow, *To Heal a Nation: The Vietnam Veterans Memorial* (New York: Harper and Row, 1985), 73.
117. Cited in ibid., 85.
118. Tom Carhart, "Insulting Vietnam Vets," *New York Times*, October 24, 1981. See also Scruggs, *To Heal a Nation*, 80–84.
119. Scruggs, *To Heal a Nation*, 159.
120. Hass, *Carried to the Wall*, 63.

CHAPTER FIVE

1. William L. Hamilton, "A Fence with More Beauty, Fewer Barbs," *New York Times*, June 18, 2006, sec. 4, 14.
2. All quotations in this section are cited in Hamilton, "Fence with More Beauty."
3. John Meyer, "The Concept of Private Property and the Limits of the Environmental Imagination," *Political Theory* 37, no. 1 (2009): 117.
4. Carol M. Rose, "Possession as the Origin of Property," *University of Chicago Law Review* 52 (Winter 1985): 79.
5. Harvey Jacobs, "Claiming the Site: Evolving Social-Legal Conceptions of Ownership and Property," in *Site Matters*, ed. Carol J. Burns (New York: Routledge, 2005), 19.
6. Eric T. Freyfogle, *The Land We Share: Private Property and the Common Good* (Washington, DC: Island Press, 2003), 2.
7. Jeremy Bentham expressed much the same idea: "Property and law are born together, and die together. Before laws were made there was no property; take away laws and property ceases." Bentham, *The Theory of Legislation*, chap. 8, cited in ibid., 4.
8. Jean-Jacques Rousseau, *A Discourse upon the Origin and the Foundation of the Inequality among Mankind*, pt. 2, cited in Freyfogle, *Land We Share*, 4.
9. Aristotle, *Nicomachean Ethics* 6.5, trans. David Ross (Oxford: Oxford University Press, 1980), 142–43.
10. Martha Nussbaum, *Love's Knowledge: Essays on Philosophy and Literature* (New York: Oxford University Press, 1990), 55.
11. Michel Foucault, *Discipline and Punish: The Birth of the Prison* (1977; repr., New York: Vintage Books, 1995), 214.
12. On the use of drones along the United States border with Mexico, see William Booth, "Keeping Watchful Eye on Border, but Staying Out of Sight," *Washington Post*, December 22, 2011, sec. A, 8.
13. Lewis Mumford, *The City in History* (New York: Harcourt, Brace and World), 358. More recently, W. G. Sebald wrote about this same transition in the novel *Austerlitz*: "It is amazing . . . to see the persistence with which generations of masters of the art of military architecture, for all their undoubtedly outstanding gifts, clung to what we can easily see today was

a fundamentally wrong-headed idea: the notion that by designing an ideal tracé with blunt bastions and ravelins projecting well beyond it, allowing the cannon of the fortress to cover the entire operational area outside the walls, you could make a city as secure as anything in the world can ever be." W. G. Sebald, *Austerlitz* (London: Penguin Books, 2002), 17.

14. See Lewis Coser, *The Functions of Social Conflict* (Glencoe, IL: Free Press, 1956).
15. This is precisely the argument made by the occupying American authorities, who ordered the construction, virtually overnight, of a concrete wall three miles long and twelve feet high—the "Great Wall of Adhamiya"—between the Sunni and Shia neighborhoods of Baghdad in 2007. See Ewen MacAskill, "Latest US Solution to Iraq's Civil War: A Three-Mile Wall," *Guardian*, April 21, 2007, Home, 1.
16. Clive Barnett, "Hospitality and the Acknowledgement of Otherness," *Progress in Human Geography* 29, no. 5 (2005): 16.
17. Dan Barry, "A Natural Treasure That May End Up without a Country," *New York Times*, April 7, 2008, sec. A, 14.
18. Ernest Pollard et al., *Hedges* (New York: Taplinger, 1974), 119.
19. Richard T. T. Forman, *Land Mosaics: The Ecology of Landscapes and Regions* (Cambridge: Cambridge University Press, 1995), 195.
20. Ibid., 85.
21. Aristotle, *Ethics* 2.6, 38.
22. Richard Sennett, *The Craftsman* (New Haven, CT: Yale University Press, 2008), 295.
23. Nicholas Blomley, *Unsettling the City: Urban Land and the Politics of Property* (New York: Routledge, 2004), 77.
24. Meyer, "Concept of Private Property," 113.
25. See David M. Smith, "How Far Should We Care? On the Spatial Scope of Beneficence," *Progress in Human Geography* 22, no. 1 (1998): 15–38.
26. See Emmanuel Lévinas, *Totality and Infinity: An Essay on Exteriority* (Pittsburgh, PA: Duquesne University Press, 1969).
27. John Rawls, *A Theory of Justice* (Cambridge, MA: Harvard University Press, 1971), 12.

Index

Note: Page numbers in italics indicate figures.

Abu Dis neighborhood of Jerusalem, 12, 118–20, *119*, 123–24, 161

Africa: ecological projects across national boundaries in, 126–27, 165; "good fences make good neighbors" and, 2; Great Green Wall of Sahara Desert and, 126–27, 165; legal and physical walls/boundaries in, 10–12, 127; Sahara Forest Project in, 126–27; walls/boundaries and, 10–12, 126–27, 165

agrarian republic ideal, 84, 86–87, 89, 93–94, 198n71

Alexander the Great, 33–34

Allen, Robert C., 79, 197n52

American West: barbed wire and, 88–96, *90*, *93*, 161, 164, 200nn122–23; capital in, 88, 94; cattle ranchers vs. homesteaders and, 91–93; Desert Land Act of 1876 and, 88; ditches and, 88; enclosure/s and, 88–89, 199n93; environmental control in, 96, 161; fences in context of property and sovereignty and, 86; fence wars and, 92–93, *93*; freehold farmers/freeholder in context of enclosure and, 91, 202n76; grid pattern and, 83–86; homesteaders and, 87–90, 88–89, 91–93, 199n93; improvement of land in, 82–83, 87–90, 94, 111–12; indigenous populations/Indians or "non-enclosers" and, 82, 86, 93–94, 165; labor and commons in context of, 88; land allotment and, 83–86; migrants in context of commons and, 86–89; moral values and, 82–83; national enclosure in context of sovereignty and, 82–83, 86, 93–94; natural ecology and, 93, 165; nature as commons or "God's great common" and, 87; New Military Tract of 1782 and, 83; Northwest Ordinances and, 84–87, *85*; settlement and travel patterns and, 91–93; social ecology and, 82, 93, 165; social hierarchies and, 105; sovereignty in context of capital and, 88, 94; speculation in land and, 88–89, 94, 102, 199n95; squatters as impediments to speculation and, 87, 91–92; surveyors and, 86–87; water holes and, 91–92. *See also* United States and history of modern landscapes

ancient maps (terriers), 54–55, 76, 145

appearance of wealth and ease, and landscape design, 100–101, 103, 106, 111–12

Apulia site, 189n20
archaeological sites: Bronze Age, 28–29; Iron Age, 33, 166; Neolithic societies, 24–30, *26*, 38, 45, 47, 166, 189n20. *See also specific sites*
arete ("virtue"), in context of walls/boundaries, 14–17, 167–68
Aristotle, 14, 16–17, 158, 167–69
Artists Without Walls, 118–19, 161
Aspen Farms, West Philadelphia, PA, 141–42, 143–44, 166–67
Assyria, 30, 32, 50
Aurelian Wall, 51, 60, 62

Babylon, 34–35, *35*, 37, 43–44, *44*, 50, 59–60, *61*, 63–65. *See also* Mesopotamia
bad walls, xx, 14, 17, 157, 170; contestation and, 157; reconciliation in context of, 8–9, 15, 18–19, 187n19
barbarism or tyranny of Old World, and walls as symbols, 7, 17, 81, 102, 105, 111–13, 201n152
barbed wire, 94–95, 200nn122–23; American West and, 88–96, *90*, *93*, 161, 164, 200nn122–23; Berlin Wall and, 115–17; cattle ranchers vs. homesteaders and, 91–93; commons theft in context of, 94; ease of use in, 90, 94; environmental control and, 96, 161; fences between countries and, 2, 10–11; fence wars and, 66–67, 92–93, *93*; improvement of land in context of, 89–90; indigenous populations/Indians or "non-enclosers" in context of, 93–94; manufacture of, 89–90, *90*; settlement and travel patterns and, 91–93; social ecology and, 164; sovereignty in context of capital and, 94; use of, 94–95, 200nn122–23; U.S.-Mexico border fence and, 2
barriers of separation (separation barriers). *See* legal and physical walls/boundaries; separation barriers (barriers of separation)
beating the bounds (boundary beating), 56, 144–47, *146*
"beautiful/noble/fine" (*kalon*), 167–68
belief: *both/and* concept and, 65; cosmos in context of walls and, 43, *44*; membrane, between living and spirit worlds and, 38–39, *39*, 152, 166; nurture of walls and, 24–30, *26*, 43; paradise wall or walled palace and, 39–41, *41*, 152; recovered purposes of walls and, 120, 153; social order in context of dwelling and, 45–47, *46*, 192n77. *See also* religious experience; ritual; symbols, walls as
Beltane festival, 55
Bentham, Jeremy, 208n7
Benton, Thomas Hart, 87, 90
Berlin, Germany: Berlin Wall and, xx, 114–18, *117*, 120, 161; perimeter block in, 137–38
"between" as separation/connection, 6, 9, 187n19
Bible (Old Testament), 17, 20–24, 50, 142. *See also* Jericho settlement and walls
binding and bounding, similarity of, 47, 96, 166, 172
Blenheim Palace, 95–96, 112
borders, international, 2, 10–12, 127. *See also specific international borders*
Bororo villages, 45–47, *46*, 192n77
both/and vs. *either/or* concept, 65, 112–13, 140
boundaries/walls. *See* walls/boundaries
boundary beating (beating the bounds), 56, 144–47, *146*
boundary stones, 34–37, *35*, 53–54, 65, 144–45
bounding and binding, similarity of, 47, 96, 166, 172
Bradford, John, 189n20
Brandenburg Gate, 114. *See also* Berlin, Germany
Brecht, Bertolt, 128, 130
Bremner, Lindsay, 12
Brezhnev, Leonid, 116
Bronze Age, 28–29
Brown, Lancelot "Capability," 95
Brown, Wendy, 10
"Byker Wall," 134–37, *136*

Cadbury Castle, 33
Cambridge, MA, 142–43
capital, and sovereignty in U.S., 88, 94
Caro, Alberto, 120
Castells, Manuel, 10
Çatalhöyük site, 37–39, *39*, 43, 152, 167
cattle ranchers vs. homesteaders, in U.S., 91–93
Celtic societies, 55

Certeau, Michel de, 123
chain-link fences, 10, 104–5, 141–44; between countries, 10; crafts and, 167–68; exchange and, 162–63; as nurture, 141–42, 143–44, 167; visibility/invisibility of, 104–5
Chaucer, Geoffrey, 64
Childe, Gordon, 25–27, 38
China, 2, 17, 33, 50–51, 155, 191n40, 193n84
Christians, 54, 119
Christo, 130–32, *131*, 165
citadel, fortified religious (*temenos*), 41, 50
city as wall (fusion of city and wall), 47–48
Clare, John, 79–81, 111
Clark, William, 86–87
Coalición pro Defensa del Migrante, 120
Cold War, xx, 115, 117. *See also* Berlin, Germany
"collective form," 132–33
commoners, as impediments to landlords, 75, 79–81, *80*, 87, 111
common law, 30, 122
common (public) good, 72–73
commons: as becoming property, 67, 69–70, 72, 74–75, 78–79, 81, 87, 111–12; labor in context of, 69, 88, 156; loss in U.S. of, 5; as marginal "waste grounds" and, 72, 74–75, 78–79; migrants in U.S. in context of, 86–89; "Ramble" in U.S. and, 105–6; as resources, 79; visibility/invisibility in British landscape design in context of enclosed, 98, *99*, 100. *See also* property
"community wall," 133–34
concrete panels/slabs, 11, 114–15, 118–19, *119*, 161–63
contestation, and ethical enclosure, 156–57, 161–62, 168, 208n13
contrast/distinction, and boundaries, 13–14
Copenhagen, Denmark, xv, 137–38, *139*
Corner, James, 154–55, 163
Cornwall, England, *28*, 28–29, 29–30, 33, 75, 124, 166
Coser, Lewis, 163
Countryside and Rights of Way Act of 2000, 122
crafts, and ethical enclosure, 167–69
Crawford, Margaret, 140
curtain walls, 43, 57, 161
"Cyclopean" (Mycenaen) masonry, 43, 57–58, *59*
Cyrus the Great, 33
Cyrus the Younger, 39–40

Davis, Alexander Jackson, 105, 107
Davis, Mollie E. Moore, 91–92
"defamiliarization" (*Verfremdungseffekt*), 128, 130, 132
defensive walls, against violence , 20–24, *23*, 43, 57–58, *59*, 189n20. *See also* nondefensive walls
democracy, xv, 10, 14, 84, 86, 111–12
Denmark, xiv–xvi, xviii, xx, 124
Desert Land Act of 1876, 88
Devon county, England, 29–30, 56, 146
differentiation, 15, 18
ditches, xx, 9, 64–65, 144, 147; American West and, 88; boundary beating and, 146–47; Britain's history of modern landscape and, 72–73; ha-ha device and, 95–96, 100, 112, 161; modern landscape in historical context and, 72–73, 88, 189n20; ritual rebuilt and, 144, 146–47; Roman Empire and, 52–53; urban civilization and, 33–34; walls' history and, 22, 34, 54–55, 189n20
Ditlevsen, Tove, 138
Downing, Andrew Jackson, 101–8, *104*, 111–12, 140, 201n145, 201n152
dwelling, and history of walls, 45–52, 192n77. *See also* walls as dwelling (living in or on walls)
Dymock, Cressey, 71–74, *73*, 77, 82–83

East Berlin. *See* Berlin, Germany
East Los Angeles, CA, and reimagined lots, 139–41, *141*, 167
ecology, 13–14; ecotone and, 166; environmental control and, 96, 100; ethical enclosure and, 164–66; national boundaries and, 126–27, 165; *oikos* defined and, 164; recovered purposes of walls and, 124–27, 137, 165; social ecology and, 50, 82, 93, 131–32, 137, 165. *See also* ethical enclosure
economics and economic exchange: "suburbs" and, 60–64, *61*, *63*; visibility/invisibility in landscape design and, 95–96, 100, 106–7, *107*, 112, 161
ecotone, 166

Egypt, 22, 34, 75, 85, 144
either/or vs. *both/and* concept, 65, 112–13, 140
Eliade, Mircea, 40, 42
Ellwood, Isaac, 89
Emerson, Ralph Waldo, 2
"enclose no land" moral justification, 93
enclosure/s: American West and, 82–83, 86, 88–89, 93–94, 199n93; hedges and stiles in context of, *28*, 28–29, 75; public or common good and, 72–73; sovereignty in context of , 82–83, 86, 93–94; suburban U.S. landscape in context of, xv–xvii, *xvii*; wattle woven fence, xvi–xvii, *xvii*, 40, 168; yeomen in context of, 79, 91, 197n52. *See also* parliamentary enclosure
England. *See* Great Britain; Great Britain and history of modern landscapes; visibility/invisibility of walls in British landscape design
environmental control, 96, 100. *See also* ecology; visibility/invisibility of walls and power
"environmental wall," 133–34. *See also* Maki, Fumihiko
Erskine, Ralph, 134–35
ethical enclosure, xx, xx–xxi, 169–71; contestation and, 156–57, 161–62, 168, 208n13; crafts and, 167–69; ecotone and, 166; *ethos* defined and, 158; exchange and, 162–64; Frost and, 158–59; gated communities and, 157, 160; "inside" and "outside" as interwoven in context of, 171–72; justifiable/unjustifiable walls and, 157, 159–60, 162, 163, 171–72, 209n15; nurture and, 166–67, 168; peace and, 163; performance in context of walls as, xix, 160, 163; reconciliation in context of, 9; recovered purposes of walls and, 163–64; ritual and, 163–64; social ecology and, 164–65; urban redevelopment projects and, xvii–xix; "virtue" in context of, 167–68; wattle fences in context of, xvii, 168; West Bank "seam line obstacle" and, 159, 161. *See also* ecology
ethos defined, 158
exchange: ethical enclosure and, 162–64. *See also* economics and economic exchange; social exchange

Farmer's and Planter's Almanac (Blum), 2
faubourg ("outside burg"), 62–63, *63*, 166
fences: chain link, 10, 104–5, 141–44; freehold property and, 107, 202n176; illusion of property vs., 103, 107–8, 111, 202n176, 202n178; improvement of land in context of, 81, 88–90, 111; as nurture, 141–42, 143–44, 167; property and sovereignty in context of, 71, 86, 111; as reimagined in East Los Angeles, 139–41, 167; as term of use, 29; urbanization in context of, 109, *110*, 111
fence wars, 66–67, 92–93, *93*
forgetfulness danger, xx, *16*, 168–69
foris burgus ("outside burg"), 62–63
Forman, Richard T. T., 166
fortified religious citadel (*temenos*), 41, 50
Foucault, Michel, 160
freeholders, 16, 91, 202n76
freehold property, 107
Freyfogle, Eric, 156
Friendship Park, 120–21
Frost, Robert, "Mending Wall": bad walls interpretation of, xx, 14; barbarism or tyranny of Old World in context of walls as symbols and, 7, 81, 105, 111; "between" as separation/connection, 6, 9, 187n19; crafts and, 168; ethical enclosure and, 8, 158–59, 168–69; fragility of walls and, 5–6; "function" and, 8; "good fences make good neighbors" and, 6–7; reconciliation in context of, 8–9, 187n19; reflection and questioning of walls and, 18–19; resignation to necessity of walls, 18, 47; ritual and, 7–8, 52; springtime meeting or Terminalia and, 56, 194n114
Fukuyama, Francis, 10, 187n21
function of walls: *ergon* and, 14; forgetfulness dangers and, xx, *16*, 168–69; Frost and, 8; goodness in context of, 14–15; premodern societies and, 24, 27, 42–44, 48, 189n13
fuori porta (outside the gates), 61
fusion of city and wall (city as wall), 47–48

gardenesque and "natural" irregularity, 101–8, *104*, *107*, 201n145, 201n152, 202n176, 202n178
gate/s: ethical enclosure and, 157, 160; *fuori porta*, or outside the gates, and, 61;

gated communities and, xviii, 11–12, *12*, 14, 122, 157, 160; gatehouses and, 51, 106–7, *107*; gating/gating orders and, 11, 122; history of walls and, 50–52, 59–60; *portus*, or, 62
Gavillard, Gilbert, 89
Geddes, Patrick, 203n181
Germany. *See* Berlin, Germany
Gilgamesh epic, 30–33, 41, *42*
Glidden, J. F., 89, *90*
"God's great common" (nature as commons), 69, 81–82, 87
"good fences make good neighbors," 1–3, 6–9, 21, 185n1
Great Britain: boundary beating and, 56, 144–47, *146*; Celtic societies, 55; common law in, 30, 122; Countryside and Rights of Way Act of 2000 and, 122; curtain walls and, 57, 161; ditches and, 29–30, 54–55; "fence" as term of use in, 29; gated communities and, 122; gatehouses and, 51; gates and, 51; gating orders in, 122; "good fences make good neighbors" and, 2; hedges and, 29–30, 144; hedges and stiles in, *28*, 28–29, 75, 124; hedges as nurture in, *28*, 28–30, 190n21; landmarks and ritual in, 54–56, 144–47, *146*; landmarks and ritual in context of, 54–56; nurture and, 166; open-field systems and, 74; outside walls activities and, 63–64; piecemeal enclosure and "terriers" or maps and, 54, 76–77, 145; Ramblers and, 121–23, *123*; rights-of-way and, 121–23, *123*; social exchange and, 56–57, 76; stiles and hedges in, 28, *28*, 122; territoriality in context of boundaries in, 33, 191n39; urbanization in, 108–9, *110*, 203n181; walls as dwelling in, 51. *See also* hedges in Great Britain; visibility/invisibility of walls in British landscape design
Great Britain and history of modern landscapes, 111–12; barbarism or tyranny of Old World in context of walls as symbols and, 81; boundaries in context of property in, 71–74, *73*, 77–79, 82–83, 86, 111; commoners as impediments to landlords and, 75, 79–81, *80*, 87, 111; commons as becoming property and, 67, 69–70, 72, 74, 81, 87, 111–12; commons as marginal "waste grounds" and, 72, 74–75, 78–79; ditches and, 72–73, 88, 189n20; fence in context of urbanization and, 109, *110*, 111; fences in context of property and sovereignty and, 71, 86, 111; fence wars in, 66–67; grid pattern and, 72–73, *73*, 77; improvement of land and, 74–75, 94, 111; individual sovereignty of the body and, 69, 156, 157; labor and commons in context of, 69, 88, 156; nature as commons or "God's great common" and, 69, 81–82, 87; open-field systems as impediment and, 74; property origins and, 67, 69, 196n16; public or common good in context of enclosure and, 72–73; ribbon development and, 108–9; social contract and, 68, 70–71, 84, 86, 89; social hierarchies and, 72–74, 97–98, 100; sovereignty in context of property and, 68–71; speculation in land and, 108–9; state of nature and, 68–70, 89; terriers, or ancient maps and, 54–55, 76, 145; urbanization and, 108–9, *110*, 203n181; yeomen in context of enclosure and, 79, 91, 197n52. *See also* Great Britain; parliamentary enclosure; visibility/invisibility of walls in British landscape design
Great Green Wall of Sahara Desert, 126–27, 165
Great Wall (Long Wall/City) of China, 33, 50, 155, 191n40, 193n84
Greek culture and cities: contrast/distinction in context of myths and, 13; curtain walls and, 43; defensive walls against violence and, 43, 57–58, *59*; *ethos* defined and, 158; *oikos* defined and, 164; *paradeisos* defined and, 40; posterns and, 58–59, *59*; *temenos*, or fortified religious citadel, 41; wall construction and, 43, 57–58, *59*; walls as symbols and, 47, 57, 194n116
grid pattern: Great Britain and, 72–73, *73*, 77; North American colonies and, 83–86; parliamentary enclosure and, 77, *77*; Roman Empire and, 54; U.S. and, 83–86, 102
Gruner, Silvia, 120

Hadrian's Wall, 33, 127–28, 127–30, *129*, 130, 155, 191n39
Haggard, H. Rider, 78

ha-ha device, 95–96, 100, 112, 161. *See also* ditches
Harbison, John, 142–43, 163–64, 167–68
Harvey, Alfred, 62
Haskell, Llewellyn, 105–6
Hass, Kristin Ann, 152
hawthorn hedges, 78, 124, 167
healing, and recovered purpose of walls, 150
hedges: boundary beating in context of, 144, 147; in Denmark, xx; ecology and, *28*; hedge networks and, 124–27, 165; in Japan, 124, 132–34, 190n26; natural ecology and, 124–27, 165, 166; as nurture, 167; social ecology and, 164; in U.S., 2, 124, 142
hedges in Great Britain, 124; as boundaries, 125–26; common law and, 30; ecology and, 124–26, 127, 190n26; events/competitions and, 125; hawthorn and, 78, 124, 167; "hedgebote" law and, 30; hedgerows and, 30, 125; "hedging" defined and, 29; as nurture, *28*, 28–30, 167, 190n21; parliamentary enclosure and, 67, 78, *80*, 124; stiles and, 28, *28*, 122, 124; walls as symbols and, 125
Hellenistic period, 58. *See also* Greek culture and cities
Herodotus, 34, 59–60
Hindus, 147
history of walls, 64–65; *both/and* vs. *either/or* concept and, 65; boundaries in context of urban civilization and, 34–37; boundary stones and, 34–37, *35*, 53–54, 65, 144–45; city as wall or fusion of city and wall fusion and, 47–48; defensive walls against violence and, 20–24, *23*, 189n20; ditches and, 22, 33–34, 54–55, 189n20; dwelling and, 45–52, 192n77; economic exchange and, 60–64, *61*; function and, 24, 27, 189n13; gates and, 50–52, 59–60; landmarks and ritual in context of, 54–56; membrane between living and spirit worlds and, 38–39, *39*; messages and, 22, 32–33, 32–37; Mycenaen or "Cyclopean masonry" and, 43, 57–58, *59*; non-defensive walls and, 24–30, *26*, 32, 38, 189n20; nurture and, 24–30, *26*, 43, 166; outside walls activities and, 33–34, 59–61; paradise wall or walled palace and, 39–41, *41*, 152; power of urban civilization and, 30–37, *35*, 42, 50; premodern societies in context of function, 24, 42–44; processioning and, 55–56, 194n109; ritual and, 22, 24, 51–56, 52–56; social exchange and, 20–22, 56–57, 56–59, 57–58, *59*, 76; temple walls and, 41–45, *42*, *44*, 50, 152, 167; territoriality and, 32–34, 191n39; walls as dwelling and, 20–22, 47–48, *49*, 50, 64, 166; walls as symbols and, 24, 27, 29, 32, 47, 57, 191n34, 194n116. *See also* belief; religious experience
Hobbes, Thomas, 68, 89
Homestead Act of 1862, 87–90
homesteaders, 87–90, 88–89, 91–93, 199n93
Honecker, Erich, 115, 116
Hoskins, W. G., 146
Hudson River school, 101–2, 201n145
Hudson River valley, 101–2, 201n145
Hunt, John Dixon, 73–74
Hunt, William D., 89

improvement of land: American West and, 82–83, 87–90, 94, 111–12; barbed wire in context of, 89–90; Britain's history of modern landscape and, 74–75, 94, 111; property and, 74, 94, 111; visibility/invisibility in U.S. landscape design and, 106
Incas, 43
India, 2, 10–11, 18, 22, 147–48, *149*, 150, 163, 207n102
indigenous populations/Indians ("non-enclosers"), 82, 86, 93–94, 165
individual sovereignty of the body, 69, 156, 157
industrialization, 101–2, 109, 112, 125, 137. *See also* premodern societies
inhabited walls. *See* walls as dwelling (living in or on walls)
"inside" and "outside," as interwoven, 18, 133, 142–44, 171–72
international borders, 2, 10–12, 127. *See also specific international borders*
Iraq, 30, 34, 43, 163, 209n15
Iron Age, 33, 166
Ishtar Gate, 50, 60, *61*
Israel: Abu Dis neighborhood of Jerusalem and, 12, 118–20, *119*, 161; concrete panels/slabs and, 11, 118–19, *119*,

161; ethical enclosure and, 159, 161; Palestinians and, xx, 11, 15–16, 118–19; political movements in context of walls and, 123–24; separation barriers and, 118, 161; sovereignty and, 11, 15–16, 21; West Bank "seam line obstacle" in, xx, 11, 12, 15–16, 118–20, *119*, 159, 161; Western Wall in Jerusalem and, 43, 144, 152, 167. *See also* Jericho settlement and walls

Italy: outside walls activities and, 64; perimeter block in, 137; walls as dwelling in, 48, *49*, 50–51

Jacobs, Harvey, 156

Jacobs, Jane, 137

Japan, 124, 132–34

Jeanne-Claude, 130–32, *131*, 165

Jefferson, Thomas, 83–90, 93–94, 108, 111, 198n71

Jericho settlement and walls: cities and, 32, 37, 189n10; defensive walls against violence and, 20–24, *23*, 27; ditches and, 22; function in context of, 24, 189n13; nurture and, 64–65; perambulation in context of, 20–21, 56; purpose of, 23–24, 189n13; Rahab as living on, 20–22, 47–48, 50, 64, 166; ritual and, 22, 24; size and history of, 22–24, *23*, 29, 189n10; social exchange and, 20–22; violence in context of sovereignty and, 21; walls as symbols and, 24

Jews, 43, 119, 144, 167. *See also* Israel

Joshua and book of Joshua, 20–22, 50. *See also* Jericho settlement and walls

justifiable/unjustifiable walls: ethical enclosure and, 157, 159–60, 162, 163, 171–72, 209n15; property and, 15–16; segregation and, 15, 163, 209n15; sovereignty and, 15–16; U.S.-Mexico border fence and, 1–2, 15; virtue/function and, 14–17, 167–68

kalon ("beautiful/noble/fine"), 167–68

karum ("suburb"), 60–61, 64

Kenyon, Kathleen, 22–23

Khan, Yasmin, 148

Khorsabad site, 50

Knight, Richard Payne, 100

Knox, Vicesimus, 2

Koldewey, Robert, 60, 63

Korean Peninsula, 11, 161, 187n27

Kosovo ethnic conflict, 2, 11

Kratsman, Miki, 118, 120

kuduru (boundary stone), 34–35, *35*, 37

labor, 69, 81, 88, 156

land allotment, in U.S., 83–86

landmarks, and ritual, 54–56

landscape boundaries, and *Running Fence* (Christo and Jeanne-Claude), 130–32, *131*, 165

landscapes and landscape design history, 38, 111–13. *See also* Great Britain and history of modern landscapes; parliamentary enclosure; United States and history of modern landscapes; visibility/invisibility of walls in British landscape design; visibility/invisibility of walls in U.S. landscape design

land surveys and surveyors, 52–54, 76–77, 86–87

legal and physical walls/boundaries, 10–12; in Africa, 10–12, 127; U.S.-Mexico border fence and, 10. *See also* separation barriers (barriers of separation)

Lévi-Strauss, Claude, 45–46, 192n77

Lewis, Meriwether, 86–87

Lewis-Williams, David, 24, 189n13

Lin, Maya Ying, 151, *153*

Lincoln, Abraham, 87

Lincoln, Levi, 84

line of boundary stones (*pomoerium*), 144

living in or on walls (walls as dwelling). *See* dwelling, and history of walls; walls as dwelling (living in or on walls)

"living" national memorials, 152

Llewellyn Park, NJ, 105–7, *107*, 202n176, 202n178

Locke, John, and topics: commons and labor, 69, 88, 156; commons as becoming property, 69–70, 74, 81, 87, 111–12; fences in context of property and sovereignty and, 71, 86, 111; improvement of land and property, 94, 111; individual sovereignty of the body, 69, 155–56; labor, 69, 88, 156; nature as commons or "God's great common," 69, 82, 87; property origins, 69, 156; social contract, 68, 70–71, 84, 86, 89; sovereignty in context of property, 68–71, 155–56; state of nature, 68–70

Long Wall/City (Great Wall) of China, 33, 50, 155, 191n40, 193n84
lots, as reimagined, *141*
lots in East Los Angeles, as reimagined, 139–41, 167
Loudon, John Claudius, 100–103, 111
Louisiana Territory, 83–84
Louria-Hayon, Adi, 118
Lupercalia festival, 54–55, 144

machiya (traditional storefront in Japan), 134
MacKaye, Benton, 203n181
Maki, Fumihiko, 132–34
Manzoni, Alessandro, 64
Marcuse, Peter, xv, xx
Marens, Richard, 71
Marshall, William, 74
masks, walls as, 143–44
Massingham, H. J., 109
McLuhan, Marshall, 9–10
medieval period (Middle Ages). *See* Middle Ages (medieval period)
"megacities," 17–18. *See also* urban civilization and urbanization
Mehdi, Syed Sikander, 150
Mellaart, James, 37–38
membrane between living and spirit worlds, 38–39, *39*, 151, 152, 166
"Mending Wall" (Frost). *See* Frost, Robert, "Mending Wall"
Mesopotamia, 22, 30–36, *31*, *35*, 41–42, *44*, 50–51, 59–60, *61*. *See also* Babylon; Persia
messages, walls/boundaries as: power in context of, 32, 33, 37, 112; recovered purpose of walls and, 119; religious experience and, 33–35, *35*, 37; resistance against, 118–24, *119*, *121*, 148, *149*, 150; topography and, 36–37; violence/warfare and, 22. *See also* United States–Mexico border fence
Mexico, and resistance against message of border fence, 119–21, *120*, 123–24, 161. *See also* United States–Mexico border fence
Meyer, John, 156
Michelangelo, 168
Middle Ages (medieval period): curtain walls during, 57, 161; ditches and, 54–55; gates and, 51; "good fences make good neighbors" and, 2; hedges and, 29–30, 144; landmarks and ritual in context of, 54–56; open-field systems during, 74; outside walls activities and, 63–64; piecemeal enclosure and "terriers" or maps during, 54, 76–77, 145; social exchange and, 56–57, 76
middle class, and landscape design in Britain, 97–98, 100
migrants, in context of commons in U.S., 86–89
modern landscapes history, 38, 111–13. *See also* Great Britain and history of modern landscapes; parliamentary enclosure; United States and history of modern landscapes; visibility/invisibility of walls in U.S. landscape design
moral values: American West and, 82–83; visibility/invisibility in landscape design and, 97, 102–3, 105
More, Thomas, 75, 78–79, 92, 109, 112
Morris, A. E. J., 63
Moss, Eric Owen, 154
Mumford, Lewis, 32, 43, 63
Muslims, 119, 147, 209n15
Mycenaen ("Cyclopean") masonry, 43, 57–58, *59*

Napoleon and Napoleonic Wars, 75, 83
"narrow cells" in U.S. cities, 102, 112
national aesthetic, and U.S. landscape design, 101–5
national boundaries, and ecological projects, 126–27, 165
national enclosure, in context of sovereignty in U.S., 82–83, 86, 93–94
national walls, 11, *11*, 121, 154–55
natural ecology: American West and, 93, 165; Great Green Wall of Sahara Desert, 126–27, 165; hedges and, *28*, 28–30, 124–27, 126, 165, 166; nurture and, 141–42, 166–67; parliamentary enclosure and, 165
nature as commons ("God's great common"), 69, 81–82, 87
Neolithic societies, 24–30, *26*, 38, 45, 47, 166, 189n20. *See also specific archaeological sites*
Newcastle, England: Hadrian's Wall and, 33, 127–30, *129*; walls as dwelling in urban redevelopment project and, 134–37, *136*
New England, 2, *4*, 5, 82–83
Newman, Oscar, 11–12
New Military Tract of 1782, 83

New Testament, 21, 40. *See also* Old Testament (Bible)
Nogales city, 120
non-defensive walls, 24–30, *26*, 32, 38, 189n20. *See also* defensive walls, against violence
"non-enclosers" (indigenous populations/ Indians), 82, 86, 93–94, 165
Norten, Enrique, 154
North American colonies, and history of modern landscapes, 70, 81–86, 93–94; commons as becoming property and, 70, 81; grid pattern and, 83–86; national enclosure in context of sovereignty and, 82–83, 86, 93–94
Northwest Ordinances, 84–87, *85*
Nozick, Robert, 196n16
nurture: ethical enclosure and, 166–68; fences/chain-link fences as, 141–42, 143–44, 167; history of walls and, 24–30, *26*, 43, 64–65, 166; natural ecology, 141–42, 166–67; property in context of, 166; U.S.-Mexico border fence in context of, 155
Nussbaum, Martha, 158

Ogelthorpe, James, 83
Old Testament (Bible), 17, 20–24, 50, 142. *See also* Jericho settlement and walls
Olmsted, Frederic Law, 107, 202n176
open-field systems, 74
Open Spaces Society, 122
"original position," 171
Orkney Islands, Scotland, 24–29, *26*, 32, 38
Otmoor, and parliamentary enclosure, 66–67, 74–75, 81, 109
"outside" and "inside," as interwoven, 18, 133, 142–44, 171–72
"outside burg": *faubourg*, 62–63, *63*, 166; *foris burgus*, 62–63
outside the gates (*fuori porta*), 61
outside walls activities, 33–34, 59–61, 63–64
Ovid, 53
"ownership" model, of walls, 16, 170. *See also* individual sovereignty of the body; property; sovereignty

Pakistan-India boundary, 10–11, 147–48, *149*, 150, 163, 207n102
Palestinians, xx, 11, 15–16, 118–19. *See also* Israel
paradeisos defined, 40
paradise wall (walled palace), 39–41, *41*, 152
Paris, France, and gates, 51–52
parliamentary enclosure: boundaries and, 77–79; grid pattern and, 77, *77*; hedges and, 67, 78, *80*, 124; improvement of land and, 75, 81; landlords' power and, 79–81, 94, 161; natural ecology and, 165; Otmoor's opposition to, 66–67, 74–75, 81, 109; process of, 76–77, *77*; "single fence"/More's lament and, 75, 78–79, 92, 109, 111–12; social ecology and, 164; social exchange in context of walls, 76, 79; surveyors and, 76–77; visibility/invisibility of walls and, 161; wage labor in context of, 81. *See also* Great Britain and history of modern landscapes
peace, and recovered purpose of walls, 150, 163
Pearce, David, 24, 189n13
performance, and walls, xix, xx, 14, 51, 130, 160, 163
perimeter block, and recovered purpose of walls, xv, 137–39, *139*
permeability/impermeability of walls, xviii, 58–59, *59*, 122, 133, 162–63
Persia, 33, 39–40, *41*, 58. *See also* Mesopotamia
physical and legal boundaries. *See* legal and physical walls/boundaries; separation barriers (barriers of separation)
Piaget, Jean, 13
Picts, 33, 127–30, *129*, 155, 191n39
"picturesque" landscape aesthetic ("pleasure garden"), 95–97, 100, 103
Pirenne, Henri, 62
place, in context of recovered purpose of walls, 146–47
Plato, vi, 13, 47, 57, 158, 194n116
"pleasure garden" ("picturesque" landscape aesthetic), 95–97, 100, 103
Plutarch, 52, 54
politics: Denmark and, xiv–xvi; original position and, 171; U.S.-Mexico border fence and, 1–2, 123–24, 185n1; walls as symbols of oppression and, xx, 114–18, *117*; West Bank "seam line obstacle" and, 123–24
pomoerium (line of boundary stones), 144
portus (gate), 62

posterns, 58–59, *59*
power: parliamentary enclosure and, 79–81, 94, 161; urban civilization history and, 30–37, *35*, 42, 50; visibility/invisibility in landscape design and, 95–96, 106–7, *107*, 112, 161
Predock, Antoine, 154
premodern societies, 24, 27, 42–44, 48, 189n13. *See also* history of walls; industrialization
Price, Uvedale, 95, 100
processioning, 55–56, 194n109
property, 67, 69, 155–56, 196n16; boundaries in context of, 71–74, *73*, 77–79, 82–83, 86, 111; commons as becoming, 67, 69–70, 72, 74–75, 78–79, 81, 87, 111–12; fences and sovereignty in context of, 71, 86, 111; improvement of land and, 74, 94, 111; individual sovereignty of the body and, 69, 156, 157; justifiable/unjustifiable walls and, 15–16; labor and, 69, 156; natural right and, 155–56; nurture and, 166; origins of, 67, 69, 155–56, 196n16; politics and, 156, 208n7; sovereignty and, 68–71
proto-landscape image, 38
public (common) good, 72–73

Radcliffe, Cyril, 147–48
Rahab, and living on wall of Jericho, 20–22, 47–48, 50, 64, 166
"Raising and Lowering the Flags" ceremony, 148, *149*, 150, 163
"Ramble," 105–6. *See also* commons
Ramblers, 121–23, *123*
Rawls, John, 171
reconciliation, in context of walls, 8–9, 15, 18–19, 187n19
recovered purposes of walls, 152–53; Abu Dis neighborhood and, 118–20, *119*, 123–24; Aspen Farms and, 141–42, 143–44, 164; belief and, 120, 153; Berlin Wall and, xx, 114–18, *117*, 120; *both/and* vs. *either/or* concept and, 140; boundary beating and, 144–47, *146*; chain-link fences as nurture and, 141–42, 143–44; "collective form" and, 132–33; common law and, 122; "community wall" and, 133–34; crafts and, 168; "defamiliarization" or *Verfremdungseffekt* and, 128, 130, 132; ecological projects and, 124–27, 137, 165; "environmental wall" and, 133–34; ethical enclosure and, 163–64; exchange and, 143, 163; Great Green Wall of Sahara Desert and, 126–27, 165; Hadrian's Wall illumination and, 127–28, *129*, 130; healing and, 150; hedge networks/hedges as ecology and, 124–27; "inside" and "outside" as interwoven, 18, 133, 142–44; "living" national memorials and, 152; lots in East Los Angles as reimagined and, 139–41, *141*, 167; and masks, walls as, 143–44; membrane between living and spirit worlds and, 151; messages of walls/boundaries in context of, 119; nurture and, 141–42, 143–44, 167; Pakistan-India boundary ritual and, 147–48, *149*, 150, 207n102; peace and, 150, 163; perimeter block and, xv, 137–39, *139*; place in context of, 146–47; political oppression symbols and, xx, 114–18, *117*; Ramblers and, 121–23, *123*; resistance against messages of walls/boundaries and, 118–24, *119*, *121*, 148, *149*, 150; rights-of-way and, 121–23, *123*; ritual rebuilding and, 144–52, *146*, *149*; *Running Fence* (Christo and Jeanne-Claude) as landscape boundary and, 130–32, *131*, 165; Sahara Forest Project and, 126–27; social ecology and, 131–32; social spaces for exchange and, 116–21, *119*, 130–32, 134, 150, 152, 165; tactic of resistance and, 121–23, *123*; territory reconsideration and, 124–32; urban block redefined and, 132–39, *139*; U.S.-Mexico border fence and, 119–21, *121*, 123–24, 154–55; Vietnam War Memorial and, 150–52, *153*, 167; *la yarda*, or space between house and street, in context of, 140–41, *141*, 164, 167
Reisner, Marc, 199n93
religious experience: messages and, 33–35, *35*, 37; *temenos*, or fortified religious citadel, and, 41, 50. *See also* belief
Renaissance, 50–51, 155, 161, 168, 208n13
Repton, Humphry, 96–103, *99*, 112
resistance, against messages of boundaries/walls, 118–24, *119*, *121*, 148, *149*, 150
ribbon development, 108–9
rights-of-way, 121–23, *123*

ritual: boundary beating and, 56, 144–47, *146*; ethical enclosure and, 163–64; exchange and, 163–64; landmarks and, 54–56, 144–47, *146*; Pakistan-India boundary and, 147–48, *149*, 150, 207n102; processioning and, 55–56, 194n109; recovered purpose of walls in context of rebuilding, 144–52, *146*, *149*; temple walls and, 41–45, *42*, *44*, 50, 152, 167; walls' history and, 22, 24, 51–56. *See also* belief
Riverside, IL, 107
Rogers, Ezekiel, 2–3, 82
Rome and Roman Empire: Aurelian Wall and, 51, 60, 62; boundary stones and, 36, 144; ditches and, 52–53; founding myth of, 52–54; *fuori porta*, or outside the gates, and, 61; gates and, 51; grid pattern and, 54; Hadrian's Wall and, 33, 127–30, *129*, 155, 191n39; Lupercalia festival and, 54–55, 144; *portus*, or gate, and, 62; ritual of *pomoerium* and, 144; separation barriers and, 33, 191n39; *suburbia* and, 60–61; surveyors and, 52–54, 76–77; Terminalia and, 53–56, 194n114; *termini*, or stone boundary markers, and, 36–37
Rousseau, Jean-Jacques, 156–57
Rudofsky, Bernard, 47–48
Running Fence (Christo and Jeanne-Claude), 130–32, *131*, 165
rural cottages of gentry/well-to-do, and landscape design, 96–97, 101–2, 106–7
Rykwert, Joseph, 52

Sack, Robert, 32
Sahara Forest Project, 126–27
Schmitt, Carl, 15
Schnorr, Michael, 120
Scott, Frank J., 104
Scruggs, Jan, 151
"seam line obstacle," and West Bank, xx, 11, 12, 15–16, 118–20, *119*, 159, 161
Sebald, W. G., 208n13
Secure Fence Act of 2006, 10. *See also* United States–Mexico border fence
segregation, 15, 163, 202n78, 209n15
Sennett, Richard, 168
separation barriers (barriers of separation): in Africa, 12; Berlin Wall and, 115, 118; "between" as separation/connection and, 6, 9, 187n19; fences in U.S. as, 11–12, *12*, 103, 107; globalization in context of, 9–10, 187n21; hedges as, 28–30, *29*; Roman Empire and, 33, 191n39; segregation and, 15, 163, 202n78, 209n15; the self in context of, 13. *See also* legal and physical walls/boundaries
settlement and travel patterns, in U.S., 91–93
Simpson, Ian, 26
Sinclair, John, 75
"single fence"/More's lament, 75, 78–79, 92, 109, 111–12
Skara Brae site, 24–29, *26*, 32, 38, 152, 166
social contract, 14, 68, 70–71, 84, 86, 89
social ecology, 50, 82, 93, 131–32, 137, 164–65
social exchange, xx; parliamentary enclosure and, 76, 79; recovered purposes of walls and, 116–21, *119*, 130–32, 134, 150, 152, 165; walls' history and, 20–22, 56–59, *59*, 76
social harmony and public order, 2–3, 106, 111
social hierarchies, 48, 72–74, 97–98, 100, 105, 138
Socrates, 171–72
Solnit, Rebecca, 12–13
sovereignty: American West and, 88, 94; capital in American West and, 88, 94; Israel and, 11, 15–16, 21; justifiable/unjustifiable walls and, 15–16; national enclosure in U.S. in context of, 82–83, 86, 93–94; property and, 68–71; U.S.-Mexico border fence and, 10–11, *11*, 15–16, 21, 155; walls as enforcement of, 15–16
space between house and street (*la yarda*), 140–41, *141*, 164, 167
"space of flows, the," 10
Sparta, 47, 57, 194n116
speculation in land: American West and, 88–89, 94, 102, 199n95; Britain's history of modern landscape and, 108–9
Spelman, Henry, 55
Spies, Werner, 131
Spirn, Anne Whiston, 141–42
springtime fertility rite (Terminalia), 53–56, 194n114
squatters, as impediments to speculation in U.S., 87, 91–92
Stands, Todd, 120

state of nature, 68–70, 89
Stenton, Frank, 62
stiles, in Britain, 28, *28*, 122
Stoltenberg, Jochim, 116–17
stone boundary markers (*termini*), 36–37
"suburb" (*karum*), 60–61, 64
suburbia, 60–61. *See also* outside walls activities
suburbs and suburban landowners: enclosure in context of, xv–xvii, *xvii*; visibility/invisibility in landscape design and, 100, 104–7, *107*
Sumer, 31–32, *42*, 166, 191n34. *See also* Gilgamesh epic
surveyors and land surveys, 52–54, 76–77, 86–87
symbols, walls as: barbarism/tyranny of Old World and, 7, 17, 81, 102, 105, 111–13, 201n152; Greek cities and, 47, 57, 194n116; hedges and, 125; historical context of, 24, 27, 29, 32, 47, 57, 191n34, 194n116; New England stone wall of lost era and, 2, *4*, 5; political oppression and, xx, 114–18, *117*; sovereignty and, 10–11, 15–16, 21, 155. *See also* belief

tactics, of resistance, 121–23, *123*
temenos (fortified religious citadel), 41, 50
temple walls, 41–45, *42*, *44*, 50, 152, 167
Terminalia (springtime fertility rite), 53–56, 194n114
termini (stone boundary markers), 36–37
terriers (ancient maps), 54–55, 76, 145
territory and territoriality: boundaries in context of, 33, 191n39; recovered purposes of walls and, 124–32; walls' history and, 32–34, 191n39
Thoreau, Henry David, 3
Tijuana, Mexico, 119–21, *120*, 123
Tilton, Theodore, 106
Timber Culture Act of 1873, 88
Toynbee, Arnold, 24, 189n13
traditional storefront in Japan (*machiya*), 134
Tuxedo Park, NY, 107
tyranny or barbarism of Old World, and walls as symbols, 7, 17, 81, 102, 105, 111–13, 201n152

United Kingdom, 2, 22. *See also* Great Britain; Great Britain and history of modern landscapes; visibility/invisibility of walls in British landscape design
United States: boundary beating in, 145–46; democratic ideals in, 14, 84, 86, 111–12; East Los Angeles lots reimagined and, 139–41, *141*, 167; gated communities in, 11–12, *12*, 14, 157; gating in, 11; Iraq invasion in 2003 by, 163, 209n15; separation barriers in context of fences and, 11–12, *12*. *See also* American West; United States–Mexico border fence
United States and history of modern landscapes, 111–12; agrarian republic ideal and, 84, 86–87, 89, 93–94, 198n71; barbarism or tyranny of Old World in context of walls as symbols and, 105, 111–13; commons as becoming property and, 70, 81; democratic ideals and, 84, 86; urban grid pattern and, 102. *See also* American West; visibility/invisibility of walls in U.S. landscape design
United States–Mexico border fence: barbed wire and, 2; contestation and, 157, 161; crafts in context of, 168; ethical enclosure and, 1–2, 154, 159–60, 185n1; exchange and, 163; "good fences make good neighbors" and, 1–2, 185n1; justifiable/unjustifiable walls, 1–2, 15, 185n1; legal boundaries and, 10; as national wall, 11, *11*, 121, 154–55; nurture and, 155; politics and, 1–2, 123–24, 185n1; procedural boundaries and, 10; recovered purposes of walls and, 119–21, *121*, 123–24, 154–55; resistance against messages of, 119–21, *121*, 123–24; Secure Fence Act of 2006 and, 10; social spaces for exchange and, 119–21, *121*; sovereignty of U.S. and, 10, *11*, 15, 155; "tactics" and, 123. *See also* United States
urban civilization and urbanization: block redefined and, 132–39, *139*; boundaries and, 34–37; in Britain, 108–9, *110*, 203n181; Britain's history of modern landscape and, 108–9, *110*, 203n181; ditches and, 33–34; ethical enclosure and, xvii–xix; gated communities and, xviii, 14, 122, 157, 160; grid pattern in U.S. and, 102; history of, 34; Japanese urban space concept and, 132–34; in modernity, 17–18; "narrow cells" in U.S.

and, 102, 112; perimeter block and, xv, 137–39, *139*; social hierarchies and, 48, 138; social spaces for exchange and, 134; urban redevelopment projects and, 134–37, *136*; urban redevelopment projects and, xvii–xix; walls as dwelling and, 134–37, *136*
Ur site, 33, 37, 41–42, 43, 152, 167, 191n34
Uruk site, 30–32, 34, 41, *42*, 43–44, 166–67
utopia, xiii–xiv, 106, 173–84

Valley Curtain (Christo and Jeanne-Claude), 130
Vanbrugh, John, and Blenheim Palace, 95–96, 112
Vancouver, Charles, 29–30
Vaux, Calvert, 105
"veil of ignorance," 171
Verfremdungseffekt ("defamiliarization"), 128, 130, 132
Vietnam War Memorial, 150–52, *153*, 167
view from and toward house, and landscape design, 96–97, 103
violence: defensive walls against, 20–24, *23*, 43, 57–58, *59*, 189n20; messages of walls/boundaries and, 22; segregation in context of, 15, 163, 202n78, 209n15; subliminal, 148, *149*, 150, 163
"virtue" (*arete*), in context of walls/boundaries, 14–17, 167–68
visibility/invisibility of walls and power: barbed wire in context of, 96, 161; ethical enclosure in context of, 96, 160–61; landowners/property relations and, 95–96, 106–7, *107*, 112, 161. *See also* visibility/invisibility of walls in British landscape design; visibility/invisibility of walls in U.S. landscape design
visibility/invisibility of walls in British landscape design: appearance of wealth and ease in context of, 100–101, 103, 112; commons as enclosed and, 98, *99*, 100; fences in context of freehold property and, 107, 202n176; fences vs. illusion of property and, 103, 107–8, 111, 202n176, 202n178; gardenesque and "natural" irregularity and, 100–101; ha-ha device and, 95–96, 100, 112, 161; middle class and, 97–98, 100; moral values of, 97; "picturesque" landscape aesthetic or "pleasure garden" and, 95–97, 100, 103; power of landowners/property relations in context of, 95–96, 100, 112, 161; rural gentry and, 96–97, 102; suburban landowners and, 100; view from and toward house and, 96–97. *See also* Great Britain and history of modern landscapes; visibility/invisibility of walls and power
visibility/invisibility of walls in U.S. landscape design: appearance of wealth and ease in context of, 102–3, 106, 111–12; barbarism or tyranny of Old World in context of walls as symbols and, 81, 102, 105, 111, 201n152; chain-link fences and, 104–5; commons or "Ramble" and, 105–6; gardenesque and "natural" irregularity in, 101–8, *104*, *107*, 201n145, 201n152, 202n176, 202n178; gatehouses and, 106–7, *107*; Hudson River school influence and, 101–2, 201n145; improvement of land and, 106; moral values of, 102–3, 105; "narrow cells" in U.S. cities and, 102, 112; national aesthetic and, 101–5; power of landowners/property relations in context of, 106–7, *107*; rural cottages for well-to-do and, 101–2, 106–7; separation barriers in context of fences and, 103, 107; suburbs and, 104–7, *107*; utopian communities and, 106; view from and toward house and, 103. *See also* visibility/invisibility of walls and power

Wagah, Pakistan, 148, *149*, 150, 163
wage labor, 81. *See also* labor
walled palace (paradise wall), 39–41, *41*, 152
walls as dwelling (living in or on walls): in Britain, 51; "Byker Wall" and, 134–37, *136*; Great Wall of China and, 155; Hadrian's Wall and, 155; Italy and, 48, *49*, 50–51; Mesopotamian palaces and, 50–51; perimeter block and, xv, 137–39, *139*; Rahab in Jericho and, 20–22, 47–48, 50, 64, 166; urban redevelopment project and, 134–37, *136*. *See also* dwelling, and history of walls
walls/boundaries, xix–xx, 12–13; city fusion with, 47–48; international borders and, 2; utopian possibilities and, xiii–xiv,

walls/boundaries (*cont.*)
173–84. *See also* ethical enclosure; Great Britain and history of modern landscapes; history of walls; landscapes and landscape design history; messages, walls/boundaries as; recovered purposes of walls; United States and history of modern landscapes
Walpole, Horace, 96
Walzer, Michael, 13
water holes, 91–92
Watt, James, 151
wattle fences, xvi–xvii, *xvii*, 40, 168
Waugh, Evelyn, 108–9
Weber, Max, 32
Weidenmann, Jacob, 104
West Bank "seam line obstacle," xx, 11, 12, 15–16, 118–20, *119*, 159, 161
West Berlin. *See* Berlin, Germany
Western European national boundaries, 10
Western Wall in Jerusalem, 43, 144, 167
Wheatley, Paul, 51
Williams, Raymond, 81
Williams-Ellis, Clough, 108–9, *110*, 203n181
Winnicott, Donald, 13
Winthrop, John, 82–83, 86, 93–94
wire cutting, and fence wars, 66–67, 92–93, *93*. *See also* barbed wire
Woolley, Leonard, 191n34

Yamagata, Susan, 120
yarda, la (space between house and street), 140–41, *141*, 164, 167
yeomen, in context of enclosure, 79, 91, 197n52
Young, Arthur, 66, 74–75, 81–82, 87, 111

ziggurat, Ur, 41–42, 191n34